A FEDERAL LEADER'S GUIDE TO CLIMATE CHANGE

POLICY, ADAPTATION, AND MITIGATION

A FEDERAL LEADER'S GUIDE TO CLIMATE CHANGE

POLICY, ADAPTATION, AND MITIGATION

FEBRUARY 2009

RACHAEL G. JONASSEN
MICHAEL E. CANES
ANNASARA D. CARNAHAN
MONICA M. DEANGELO
J. ROBERT HARDISON
JULIA A. KALLOZ
MATTHEW W. PHILIPP
LISA M. P. WATTS

Logistics Management Institute ("LMI") is a 501(c)(3) not-for-profit corporation. This publication advances LMI's not-for-profit purpose and mission.

ISBN 978-0-9661916-5-3

Library of Congress Control Number: 2008942335

Cover photo source: The NASA Earth Observatory, earthobservatory.nasa.gov/images/imagerecords/0/885/modis_wonderglobe_lrg.jpg

Printed in the United States of America.

Foreword

LMI created *A Federal Leader's Guide to Climate Change* because climate change will be one of the defining issues of this century. It has moved from the province of specialists in environmental issues to one of concern for all government leaders. LMI is a not-for-profit research organization committed to helping government leaders and managers reach decisions that make a difference on issues of national importance. This guide and our substantial in-house expertise in the field support our continuing effort to help the government prepare for this challenge.

Those who lead agencies that are directly responsible for policy and science related to climate change will be joined by leaders of other agencies that, as significant emitters of greenhouse gases, will be called to action through executive orders, regulations, and laws. Those whose formal training is far from the environmental sciences will become involved, many with great passion. This guide—with its unique focus from the perspective and on the responsibilities of the federal leader—will assist the full spectrum of federal managers as they prepare to address the issues associated with climate change.

The guide offers a well-founded, unbiased, public, and nonpartisan introduction to the way that climate change will affect the operation of the federal government. It furnishes federal leaders a succinct, understandable, yet rigorous reference that explains specific issues and actions within the federal realm. It details the basic information leaders need when executing their responsibilities and illustrates actions through specific examples.

The International Panel on Climate Change, the U.S. Global Change Research Program, and individual U.S. agencies have produced important studies of climate change. However, many findings on this complex issue are tentative and cannot as yet form the basis for action by agency leaders. Others have been so thoroughly demonstrated by the scientific method that it would be a failure of responsibility to ignore them. This latter set forms the basis for *A Federal Leader's Guide to Climate Change*.

The guide, organized in seven chapters, presents the scientific and technological basis for acting in response to laws, regulations, and executive orders. The chapters address climate policy, the science of climate change, national and regional programs to curb greenhouse gases, carbon management, impacts of climate change on federal agencies, adaptation strategies, and international implications of climate change. The guide will broaden a federal leader's perspective on climate change and facilitate successful interaction with other agencies and governments to optimize the U.S. government's response to this global challenge.

William B. Moore, PhD, PE
Vice President, LMI

Acknowledgments

We strived to make this book a practical guide for the many federal leaders who will face the enormous challenge of climate change in the coming years. In doing so, we relied on the help of many in the federal and academic sectors to concentrate our efforts. All external reviewers are listed at the end of the book, and we extend our heartfelt thanks for their voluntary efforts. We would also like to thank our scholarly researchers at the LMI Research Institute for their review, advice, and counsel. Their thought leadership was instrumental in the success of this effort.

We would particularly like to thank Roger Pielke Jr., University of Colorado at Boulder, who advised the LMI project team from its inception through completion of the effort, and Emil J. Dzuray Jr., U.S. Postal Service, who contributed significantly to the concept of this book and Chapter 6.

Further, we are especially grateful to the following federal managers for their assistance at various stages of this effort: Sam Higuchi, National Aeronautics and Space Administration; Bill Goran, U.S. Army Corps of Engineers; Tom Fish, National Park Service; A. J. Singletary, U.S. Department of Transportation; and Peter Schultz and Bob Curran, U.S. Global Change Research Program.

LMI Authors

Rachael G. Jonassen, PhD, leads the LMI climate change practice. She was program director at the National Science Foundation, leading research into the carbon cycle as the NSF contribution to the U.S. Global Change Research Program. She served on the interagency working group for carbon cycle, where she helped manage the first *State of the Carbon Cycle Report*, led the effort at international cooperation on bilateral climate change agreements, and organized the first all-investigators meeting on the carbon cycle. Dr. Jonassen also was a lead author of the *Strategic Plan for the U.S. Climate Change Science Program*. Her PhD is from The Pennsylvania State University.

Michael E. Canes, PhD, is an internationally recognized economist with an extensive background in the economics of energy and climate policy. He has published a number of studies related to climate policy and has developed a method for estimating the impacts of policy choices on greenhouse gas emissions. Dr. Canes is an officer in the United States Association for Energy Economics. His PhD in economics is from the University of California, Los Angeles.

AnnaSara D. Carnahan, MPH, is a public health specialist who focuses on changing climate impacts on the health of populations. Her MPH is from The George Washington University.

Monica M. DeAngelo, MS, is an environmental engineer with interest and expertise in utilizing renewable energy technologies to reduce environmental impacts and increase economic stability and energy security. Her MS in earth and environmental engineering is from Columbia University.

J. Robert Hardison, MBA, MPA, is an environmental engineer policy specialist with expertise in the technical, economic, and implementation realms. His master's degree in business administration and environmental policy is from the University of Washington, and he is a LEED accredited professional.

Julia A. Kalloz, BA, is an environmental scientist working on sustainability reporting and climate change issues. She is pursuing a master's degree in public policy from The George Washington University, holds an honors BA from Villanova University, and is a LEED accredited professional.

Matthew W. Philipp, MEM, is an environmental analyst who focuses on the government's balance between national security requirements and ecological responsibility. His MEM in coastal environmental management is from Duke University.

Lisa M. P. Watts, MEM, is an environmental scientist working on program risk assessment and international environmental programs. Her MEM is from Duke University.

Contents

FIGURES

TABLES

Chapter 1
Climate Change and the Federal Domain

Unequivocal evidence confirms warming of the climate system in the last 60 years, mostly due to emissions of greenhouse gases (GHGs) from human activity. Changes include increases in global average air and ocean temperatures, widespread melting of snow and ice, and rising global average sea level.[1] *Climate change is real, and it will affect your public life and official duties.*

> Most of the observed increase in global average temperatures since the mid-20th century is very likely due to the observed increase in anthropogenic greenhouse gas concentrations.[2]

Federal leaders will play a major role in addressing the challenge posed by climate change. This *Federal Leader's Guide to Climate Change* clarifies why impending climate policies are relevant to you and your organization's mission today and helps you understand climate change basics so you can set and execute policies tomorrow. Although climate change may be the most complex and daunting intergenerational issue the United States has faced, this guide aims to prepare you and your agency for action, despite the uncertainty inherent in projections.

Authoritative, documented, and thoroughly reviewed international scientific findings form the basis of the issues and options presented here. We clarify these findings with equally well-vetted U.S. government-based studies and explain their individual implications for the federal leader, including the opportunities for action and the consequences of failing to act.

This chapter describes real changes in climate observed now, the challenges climate change presents to you, and the key resources available and explains how we have selected the material in the remainder of this book.

The next chapter explains the scientific basis of these and other findings, Chapter 3 reviews proposed programs, Chapter 4 gives options for mitigation, Chapter 5 points to likely impacts, and Chapter 6 presents approaches for adaptation. Lastly, Chapter 7 reviews opportunities for international cooperation to address climate change.

HOW WE EXPLAIN CLIMATE CHANGE

Specific findings derived from the Intergovernmental Panel on Climate Change (IPCC)—the *facts* of climate change—frame the topics discussed in this guide. This panel represents the most broadly supported, systematically organized and operated, and most widely vetted resource on climate change now available to the policymaker. Comprising leading international experts nominated and vetted by their respective governments, the IPCC has released four assessment reports at approximately 5-year intervals since 1990. We draw primarily from the fourth assessment report (AR4), which appeared in 2007.

Each IPCC assessment report provides a comprehensive and critical assessment of the scientific, technical, and socioeconomic dimensions of climate change. The most recent, AR4, describes advances in knowledge since the third assessment report (2001). These publications include a summary for policymakers (SPM) focusing on policy-relevant questions using information drawn from the voluminous report.

Confidence in Knowledge

The SPM of AR4, accepted by the U.S. government,[3] constitutes the most comprehensive statement of defensible conclusions about climate change now available.[4] This guide expands upon selected key findings of the AR4 SPM in enough detail to form an understandable whole and inform your future actions. These actions will include selecting tradeoffs between mitigation and adaptation, implementing mitigation and adaptation strategies, and changing the operations of your agency in response to new challenges from climate change impacts.

In reviewing AR4, we identified more than 2,700 findings, each of which is accompanied by some measure of uncertainty, ranging from low to high. The entire IPCC report is thousands of pages long, and the list of findings is hundreds of pages long, making them nearly impenetrable and challenging to distill. To focus, we have selected only findings in which the IPCC expressed a high level of scientific understanding (see purple boxes throughout this book and Appendix C).

The IPCC assessment process assigns a well-defined level of scientific understanding to each finding, using a uniform and carefully vetted process.[1] Findings may be qualitative in nature, with the understanding expressed in qualitative terms reflecting the degree of agreement of the evidence and the amount of evidence, as illustrated below.

Degree of scientific understanding (qualitative)		**Evidence**	
		Much	Low
Agreement	**High**	***Most reliable***	
	Low		*Least reliable*

When there is high agreement and much evidence and where a quantitative assessment of confidence is possible, the IPCC expresses that confidence using the following terminology.

Confidence (quantitative)	Chance of being correct (of 10)
Very high	**At least 9**
High	About 8
Medium	About 5
Low	About 2
Very low	Less than 1

When there is high agreement and much evidence and when the information allows estimates of the probability of events (by quantitative analysis or expert judgment), the IPCC makes a probabilistic assessment using terms referring to the likelihood of the event.

Certainty (probability)	Probability of occurrence (%)
Virtually certain	**> 99**
Extremely likely	**> 95**
Very likely	**> 90**
Likely	> 66
About as likely as not	33 to 66
Unlikely	< 33
Very unlikely	**< 10**
Extremely unlikely	**< 5**
Exceptionally unlikely	**< 1**

From the criteria, some findings in AR4 are clearly quite tentative and should not form the basis of your actions today; we do not discuss the more tentative findings in this guide. **Bold** and shading in the above lists, constitute the facts of climate change best supported by the world community at this time.

Scientific Support

We supplement, clarify, and expand on the IPCC findings with related material from other sources directly relevant to the concerns of the federal leader. The U.S. Climate Change Science Program (CCSP) produces the most reliable and extensive resource for climate change information directly related to the United States. Another very reliable source is the National Research Council (NRC) of the National Academy of Sciences (NAS).

We also draw material from the following sources, and we ascribe the reliability of those sources as indicated below:

- *CCSP synthesis and assessment products (SAP).* High, thorough review process. Confidence of findings occasionally reported.
- *NRC reports.* High, thorough review process.
- *Government agency reports.* Moderate, frequently reviewed.
- *Peer reviewed literature.* Moderate to high, limited review.
- *Nongovernmental organization reports.* Low to high, represent the agenda of organization, variable review process.

In conclusion, this guide contains information about the changing climate system, ways to mitigate that change, and ways to adapt to that change. That information is framed by the bounds of well-substantiated facts about climate change, based on the most reliable source available, and approved by the U.S. government. We focus our presentation upon the concerns of the federal leader responsible for mitigating and adapting to climate change.

Climate Change Is Real

The international consensus is that the climate is changing, the change is impacting important earth and human systems, and humans are responsible for that change. These findings demand that we understand the system, formulate policies to mitigate the change, and prepare to adapt to changes that are unavoidable.

On February 2, 2007, the IPCC removed any uncertainty about whether earth's climate has been changing. They found, in the strongest terms, that the climate system is warming.

> Warming of the climate system is unequivocal.[2]

This obvious, measured warming of the climate system, and the other findings presented here, have major implications for you: the earth's climate will continue its warming trend and impact human systems.

The Changes

The global average annual temperature increased over the past century by 0.7°C,[5] and at a more rapid rate since the 1970s. The mean temperature for the contiguous United States has increased by 0.6°C since 1900.[6] From 1979 to 2003, the rate of warming in the United States has been about twice the rate of the global average. Average U.S. surface temperatures since 1979 are increasing at nearly 1°C per decade, and minimum daily temperatures have increased more than the maximum ones. In the last century, these U.S. increases differ from region to region and are largest in Alaska.[7]

> It is very likely that over the past 50 years cold days, cold nights, and frosts have become less frequent over most land areas, and hot days and hot nights have become more frequent.[2]

Other long-term changes observed include the following:

- The average land temperature is the warmest since global records began in 1880.
- In the polar regions, 2007 was the warmest year on record for the Arctic and continued a general, Arctic-wide warming trend that began in the mid-1960s.
- Measurements of the mass balance of glaciers and ice caps indicate that in most of the world, glaciers are shrinking. The duration and extent of the summer surface melt of the Greenland ice sheet both set records (see Chapter 5, Figures 5-7 and 5-8).[8]

The Human Cause

Six key facts are cause for concern:

1. Anthropogenic GHGs in the atmosphere are now very rare but very effective in warming the earth.
2. Human activity adds GHGs beyond natural levels; humans are the only sources of some kinds of GHGs.
3. The rarity of these gases means the human contribution greatly increases their amounts in the atmosphere.

4. As economies grow and produce more energy, the increase tends to accelerate each year.
5. These gases last a long time in the atmosphere—for carbon dioxide (CO_2), 100 years or more.
6. The abundance of GHGs is changing the climate.

Considered together, these facts concern scientists; educated laypersons; politicians; business leaders; federal, state, and local leaders; and international bodies. Combined with observations of significant changes in the earth's climatic, hydrologic, and biologic systems, they call for action.

Let's look at these six facts one by one.

GHGs: Rare but Effective

GHGs reduce the loss of heat into space and raise global temperatures. They are essential in maintaining the temperature of the earth: without this natural greenhouse effect, the planet would be too cold for human habitation.

Table 1-1 shows the concentrations of GHGs in the atmosphere. Because these concentrations have already increased so much since human activity began to contribute to their abundance, we give figures for the "natural" abundance—before the burning of coal and new farming practices of the industrial revolution began to add gases. We also show the so-called atmospheric lifetime of the gas.

> ***Atmospheric lifetime.*** *A measure of the time needed for the gas to be removed from the atmosphere.*

Table 1-1. GHG Characteristics

GHG name	Pre-1750 concentration	Current concentration	Atmospheric lifetime (years)	100-year warming potential
CO_2	280 ppm	384 ppm	50–200+	1
Methane (CH_4)	700 ppb	1,735-1,857 ppb	12	25
Nitrous oxide (N_2O)	270 ppb	320-321 ppb	114	298
Sulfur hexafluoride	0	6.03-6.40 ppt	3,200	22,800
Perfluorocarbons (PFCs), hydrofluorocarbons (HFCs), and other synthesized gases	0	3.2 to 541 ppt	5 to 100	146 to ~10,900

Although GHGs are trace gases in the atmosphere, measured in parts per million (ppm), parts per billion (ppb), or parts per trillion (ppt), these small concentrations regulate the global climate system. It is precisely because they are found in such small quantities that it is possible for human activities to change their concentration.

The last column in Table 1-1 lists the contribution of each gas to warming of the earth in terms of its warming potential. The warming potential is like an index of warming, with CO_2 having a value of 1, and N_2O, for example, having a value of 298, because it is 298 times more effective at warming than CO_2. This column does not take into account the fact that CO_2 is far more abundant than the other GHGs. As Table 1-1 shows, the other GHGs are actually more effective (pound for pound or, more exactly, molecule for molecule) at warming. They are increasing faster and generally have longer atmospheric lifetimes. Certain chemically synthesized GHGs have a global warming potential more than 20,000 times that of CO_2 and atmospheric lifetimes up to 3,200 years. Chapter 2 explains how these GHGs warm the earth.

Although the details on GHGs shown in Table 1-1 may be intimidating, the fact that at least 93 GHGs are known and the number is increasing as chemists create others for new applications is even more staggering. Each of these gases has distinct warming potential and lifetimes.

Human Activity

Energy use generates most of the GHGs emitted in the United States, and CO_2 is the main component of these emissions (Figure 1-1). This CO_2 comes from power generation and transportation. Cement production also adds CO_2 to the atmosphere.

Agricultural feedlots, rice farming, gas flaring, and landfills release CH_4. Use of fertilizers releases N_2O. The electrical industry uses three-quarters of all sulfur hexafluoride as a dielectric. Other uses include casting of magnesium and as an inert filling for windows. Use of refrigerants releases chlorofluorocarbons (CFCs). The gases with high global-warming potential, such as HFCs and hydrochlorofluorocarbons (HCFCs), are refrigerants and special solvents such as those used in semiconductor production.

Figure 1-1. Sources of U.S. GHG Emissions (Gt CO_2-eq)

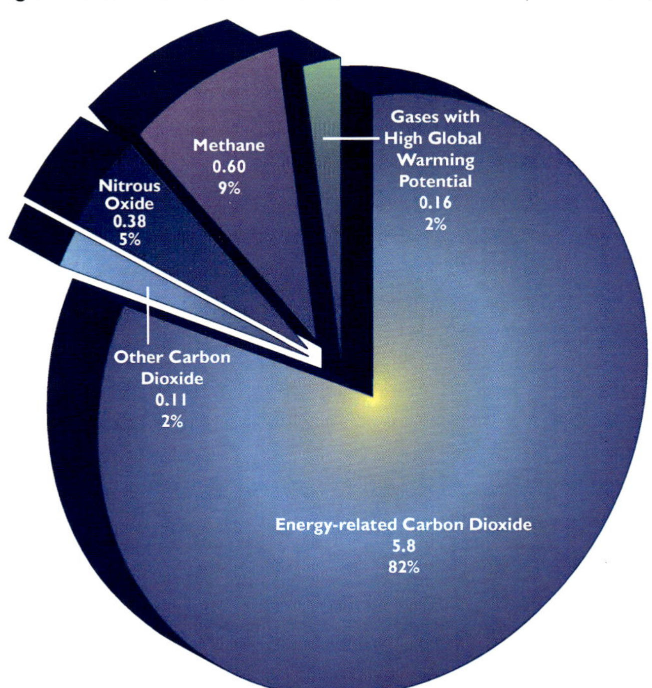

Atmospheric Concentration

Take the number given for the largest source of CO_2 in Figure 1-1, 5.8 gigatons (Gt). Totaling the other sources in Figure 1-1, we get over 7 Gt CO_2 equivalent gases released each year in the United States by human activity. Consider the amount released in 10 years: 70 Gt at the current rate of release. The atmosphere now holds about 2,163 Gt of CO_2 equivalent from natural sources (590 Gt carbon) and 686 Gt from human sources (187 Gt carbon).[9] Clearly, human emissions are changing the amount of CO_2 and other GHGs in the atmosphere very quickly.

> GHG emissions can be reported in terms of carbon or CO_2. To convert 1 Gt of carbon to CO_2, multiply by 11/3 which is the ratio of the molecular weight of CO_2 to that of carbon.
>
> Further, because climate change is international in scope, most discussions of amounts of carbon, including this guide, refer to metric tons. A metric ton is 2,200 lb (1,000 kg), whereas a short ton, commonly used in the United States, is 2,000 lb.

The primary concern is that today's human activities have increased concentrations of GHGs in the atmosphere (Table 1-1) to the point that increased global temperatures might change the climate system for future generations.

Expanding Production

Figure 1-2 shows how global emissions of carbon have increased since the beginning of the industrial age when large-scale fossil fuel use began.[10] The United States had been the largest contributor of emissions, but that has recently changed. With the growth of the Chinese economy, that country now emits more carbon than the United States, and emissions are growing faster in developing nations than in the developed world, so that by early in the next decade those countries will contribute the largest portion of total emissions.

Figure 1-2. Global Carbon Emissions from Fossil Fuel Burning, Cement Manufacture, and Gas Flaring, 1800–2004

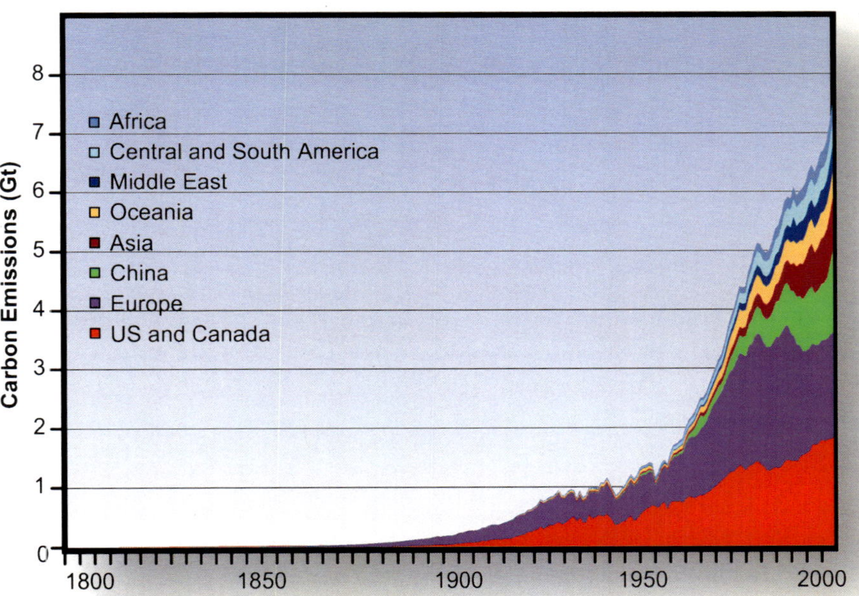

Most troubling is the acceleration in emissions growth. During the 1990s, emissions rose by about 1.1 percent per year, and the world total was about 6.8 Gt of carbon (25 Gt CO_2) in 2002, but in this decade, emissions have been growing by more than 3 percent per year. So, even keeping the rate of increase of emissions from increasing has not been possible. This only means that the total amount (and concentration) in the atmosphere will increase, and is trending to do so faster and faster.

There is high confidence and much evidence that global GHG emissions have grown since pre-industrial times, with an increase of 70 percent between 1970 and 2004.[2]

PERSISTENCE IN ATMOSPHERE

As Table 1-1 shows, each of the GHGs has a very long atmospheric lifetime. The processes that remove GHGs from the atmosphere are not the same as those that put it there. For example, rainfall removes some CO_2, but CO_2 also mixes with ocean waters, reacts in photosynthesis, and is absorbed in soils. So, estimating the atmospheric lifetime of CO_2—which can vary as these other processes vary—is difficult. Chemical reactions remove some GHGs, such as CH_4. That reaction is well known, so we can calculate the atmospheric lifetime of CH_4 quite accurately. Ultraviolet (UV) light breaks down many synthetic GHGs very slowly, so they have very long atmospheric lifetimes.

These long atmospheric lifetimes mean that GHGs now in the atmosphere will remain a problem for a long time, and this enormously complicates the effort to avoid climate change.

Stopping the acceleration in emissions growth of GHGs (the change from 1.1 percent to 3 percent growth rate) will not stop the impact of increasing emissions in the future. Even if the increase were reduced back to 1.1 percent per year or stopped completely, concentrations of GHGs would still continue to increase. The fact that we are emitting enormous amounts into the atmosphere would still be a concern and the concentration would still increase beyond that shown in Table 1-1.

A truly enormous effort would be required to reduce emissions, but that reduction in emissions would not lead to a reduction in the concentration of these GHGs. The concentration will continue to increase as long as we are emitting these GHGs. If we could bring our emissions to zero, the existing load of CO_2 (Table 1-1) still would be a concern since it vastly exceeds the natural concentration.[11]

The United States and the world are already committed to significant climate change.

Because GHGs have a long atmospheric lifetime, the concentrations would take hundreds or even thousands of years at zero emissions to reduce back to natural levels. Even if we somehow find a way to reduce the concentration in the atmosphere, it would probably take a long time, and during that time, the climate system would still be in the process of adjusting to the higher concentrations that now exist.

This is why we say that the United States and the world are already committed (irrevocably bound) to significant climate change.

Real Climate Change

There is no question that humans are responsible for emitting large quantities of GHGs into the atmosphere. Measurements demonstrate the increasing concentration of these GHGs. It is also clear that once in the atmosphere, they will remain for long periods—in most cases, beyond the human lifespan. So what? Is the change of concentration important? Is it cause for concern?

The answer to "So what?" also comes from a long-term perspective on concentrations of naturally occurring CO_2 in the atmosphere. Geologic records painstakingly extracted by researchers over decades show two things of great importance (Figure 1-3). First, the recent change in concentration vastly exceeds any natural variability that we know from this record (see the spike in the red line on the far right of Figure 1-3). Indeed, the IPCC concludes (in a "robust finding") that concentrations of CH_4 and CO_2 now "far exceed the natural range over the last 650,000 years."[12] Second, the variation over time of CO_2 closely tracks that of global temperature. Can we expect temperatures to spike just as CO_2 has? There is great concern that it will.

Figure 1-3. Long-Term Changes in CO_2 and Temperature

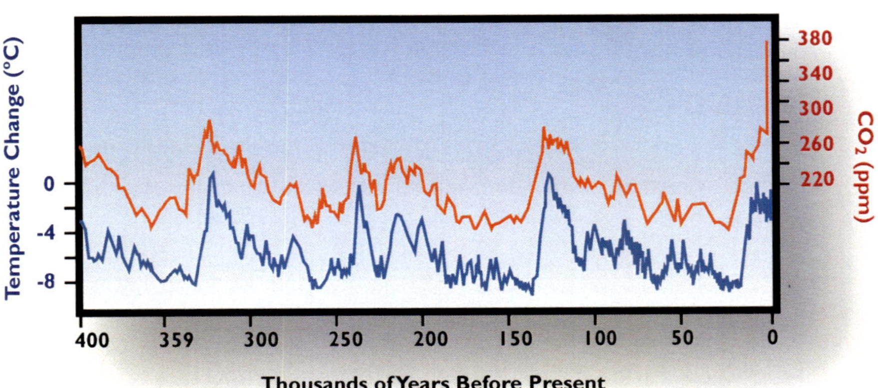

How much could temperature change? Look at the record in Figure 1-3. Over 400,000 years global average temperature varied from today's value by about 6°C, while at the same time CO_2 varied by about

100 ppm. Times of high CO_2 correspond to times of high temperature and vice versa. However, in the past 100 years, CO_2 has spiked 100 ppm above its long-term range. A simple analogy would suggest that temperatures could rise by 6°C in response. Chapter 2 reports a more careful analysis.

Think of a 6°C temperature rise in your own world. This is more than 10°F global average increase. Imagine if the temperature each day were 10°F warmer. It might seem desirable in the winter but how about during the heat of summer. Instead of 20 miserable August days at 90°F, you can imagine those same 20 days at 100°F.

As you will learn in Chapter 2, the evidence shows we face even more drastic and unprecedented global temperature changes in the future, and because of the great lifetimes of these gases in the atmosphere, we are already committed to significant changes for a long time. We don't know what is in store for the earth, but we have good reason to suspect these concentrations of GHGs pose significant problems.

Uncertainty

Prudent federal policy leaders will continue to scrutinize the climate science as it evolves, recognizing that many aspects of the climate change picture are uncertain, including the exact amount of climate change in the future and how changes will vary from region to region and over time across the globe. Some of this uncertainty concerns our ability to control our GHG emissions in the future, and some derives from incompletely understood feedbacks in the earth system (see Chapter 2).

Uncertainty also derives from increased variability of climate and increased heterogeneity of its impacts. As discussed in Chapter 5, the consensus is that the climate will vary more in the future. That variability may include more frequent changes and extreme events beyond the range of historical experience. And the impacts will vary greatly from one area to another.

Your confidence in choosing appropriate, effective, and timely adaptation strategies will increase as science narrows the uncertainties about climate change. Federal policy leaders will have to consider critical tradeoffs when deciding whether to act sooner with imperfect information or delay action in expectation of reduced uncertainties.

The IPCC finds that delaying action will increase the costs of impacts and make it more expensive to reach the same level of mitigation. As will be seen in Chapter 4, many mitigation strategies bear negative costs (that is, they pay for themselves). Thus, a cost-benefit analysis suggests immediate action on those. Even so, these easy decisions will not provide enough mitigation to prevent climate change.

YOUR ROLE IN CLIMATE CHANGE

Your wide range of responsibilities as a leader will expand again with new federal climate policies. Yet, with limited budgets and much at stake, the federal leader must critically evaluate the options available. For example, the federal government is the largest energy consumer in the United States,[13] and your decisions on energy use affect the broader economy.

Available responses to climate change are to mitigate (slow, stop, or reverse the change) and adapt (identify impacts, and change behaviors and institutions). This chapter introduces these responses—mitigation and adaptation—which are discussed in detail in Chapters 4 and 6. Your job as a leader is to select and balance between these two responses. Mitigation now may reduce the need for or facilitate adaptation in the future.

Federal Leadership

To date, addressing climate change may not have been a priority in your organization. For example, in 2007, the U.S. Government Accountability Office (GAO) acknowledged, "addressing the effects of climate change is currently not a priority within the federal resource management agencies."[14] U.S. policy is dynamic and requires active involvement by federal leaders. More change will come in light of the public interest and political climate.

Our survey of 2007 strategic plans, annual performance reports, and highest-level budget request documents for 28 federal departments shows that few report GHG emissions or plans for mitigation or adaptation (Table 1-2), although most allocate funds for climate change research. Because the cost of delay of mitigation and adaptation could be significant in terms of economic and environmental consequences for your agency and the country, early action requires your leadership to develop strategic plans and budgets.

Table 1-2. Federal Agency Reporting on Climate Change Actions and Plans

Climate change action	Agency source	Number reporting	Additional sources identified
GHG emissions	Annual performance report	0	FEMP energy data reports (3), environmental report (1)
GHG target	Annual performance report	0	Department regulation (1), environmental report (1)
GHG mitigation	Budget request or strategic plan	4	Environmental report (1), news release (1)
Current adaptation programs	Annual performance report	1	Congressional testimony (1)
Future adaptation programs	Budget request or strategic plan	3	Climate change research strategic plan (1)

Notes: Data for FY07. FEMP = Federal Energy Management Program.

Policies

In 2007, the Bush administration articulated four components of federal climate change policy:

1. Reduce the GHG intensity (total emissions/gross domestic product) of the U.S. economy.
2. Insist that mandatory GHG emissions reductions apply to all countries (for example, it makes no sense to reduce our emissions by generating more emissions elsewhere).
3. Support continued scientific study of climate change.
4. Develop emissions-reducing technologies.[15]

Options to reach the voluntary GHG intensity reduction target include

- energy efficiency and GHG reduction requirements,
- fostering energy-efficient transportation, including mass transit,
- fuel switching,
- fiscal policies that encourage conservation and renewable energy, and
- the Environmental Protection Agency (EPA) and Department of Energy (DOE) sharing federal government expertise across all agencies.

In January 2007, the Bush administration issued Executive Order (EO) 13423,[16] which requires agencies to conduct their environmental, transportation, and energy-related activities in an environmentally, economically, and fiscally sound, integrated, continuously improving, efficient, and sustainable manner. Specific targets of EO 13423 include reducing GHG intensity by 3 percent per year (which would balance the current growth rate in emissions) and obtaining at least half of the renewable energy used by federal agencies from new sources rather than existing ones. This would encourage the development of new renewable energy sources and lead to lower emissions.

In May 2007, the Bush administration issued EO 13432,[17] which directs EPA, DOE, the Department of Transportation (DOT), and the U.S. Department of Agriculture (USDA) to take the first steps toward regulations intended to cut gasoline consumption and GHG emissions from motor vehicles by reducing U.S. gasoline consumption by 20 percent over the next 10 years.

Support from U.S. Programs

Begun as a presidential initiative in 1989, the U.S. Global Change Research Program (USGCRP) supports research on the interactions of natural and human-induced changes in the global environment and their implications for society. Congress codified it in the Global Change Research Act of 1990, which mandates development of a coordinated interagency research program. The Office of Science and Technology Policy (OSTP), Office of Management and Budget (OMB), and Council on Environmental Quality (CEQ) oversee the program on behalf of the Executive Office of the President,[18] in which 13 agencies participate (Table 1-3).

The CCSP (U.S. Climate Change Science Program) and the Climate Change Technology Program (CCTP) represent the nation's efforts to apply science and technology to the problem of climate change within the structure of the USGCRP. They are multiagency research and development (R&D) coordination activities, organized under the auspices of the Committee on Climate Change Science and Technology Integration (CCCSTI). The cabinet-level CCCSTI (Figure 1-4) is chartered to provide recommendations to the president on matters concerning climate change science and technology, address related federal R&D funding issues, and coordinate with OMB on implementing its recommendations.

Table 1-3. USGCRP Agency Participation

Agency	Emphasis of Activities
USAID	The Famine Early Warning System Network; support in 40 countries for reduced growth in GHG emissions while promoting energy efficiency, forest conservation, biodiversity; partnerships with the private sector
USDA	Risks to natural resources, the role of forestry and agricultural activities in GHG emissions and carbon sequestration, and practical strategies to manage emissions and adapt to changes
DOC (NOAA, NIST)	Climate observation and monitoring, climate research and modeling, and climate services development; measurements and standards that support accurate and reliable climate observations
DoD	Satellite sensors and observations, global observations and models, polar regions research
DOE	Climate change modeling, climate forcing, and climate change response
HHS (NIH)	UV and near-UV radiation research and research related to the health effects of global change
DOS	Support to UNFCCC, IPCC, bilateral programs, and the Asia-Pacific Partnership on Clean Development and Climate, Carbon Sequestration Leadership Forum, Group on Earth Observations, Methane-to-Markets Partnership, International Partnership for a Hydrogen Economy, and Generation IV International Forum
DOT (Climate Change Center)	Potential impacts of climate variability and change on transportation infrastructure and services, energy efficiency and GHG reduction, and transportation GHG data and modeling
DOI (USGS)	Climate history and impacts on landscapes and ecosystems, hydrologic impacts of climate change, carbon cycle science, land-use and land-cover changes, and decision-support research and development
EPA	Potential consequences of global change (particularly climate variability and change) on air quality, water quality, aquatic ecosystems, and human health in the United States
NASA	Interdisciplinary science of the global integrated earth system, using 14 on-orbit satellites and aircraft- and surface-based instruments to calibrate and enhance interpretation of high-accuracy, stable satellite measurements
NSF	Fundamental understanding of physical, chemical, biological, and human systems and the interactions among them
Smithsonian Institution	Atmospheric processes, ecosystem dynamics, natural and anthropogenic environmental change on daily to decadal time scales, and longer-term climate proxies present in artifacts and records of the museums and field sites

Note: USAID = U.S. Agency for International Development; DOC = Department of Commerce; NOAA = National Oceanic and Atmospheric Administration; NIST = National Institute of Standards and Technology; DoD = Department of Defense; HHS = Department of Health and Human Services; NIH = National Institutes of Health; DOS = Department of State; UNFCCC = United Nations Framework Convention on Climate Change; DOI = Department of the Interior; USGS = U.S. Geological Survey; NASA = National Aeronautics and Space Administration; NSF = National Science Foundation.

Figure 1-4. Climate Change Policy Management Structure

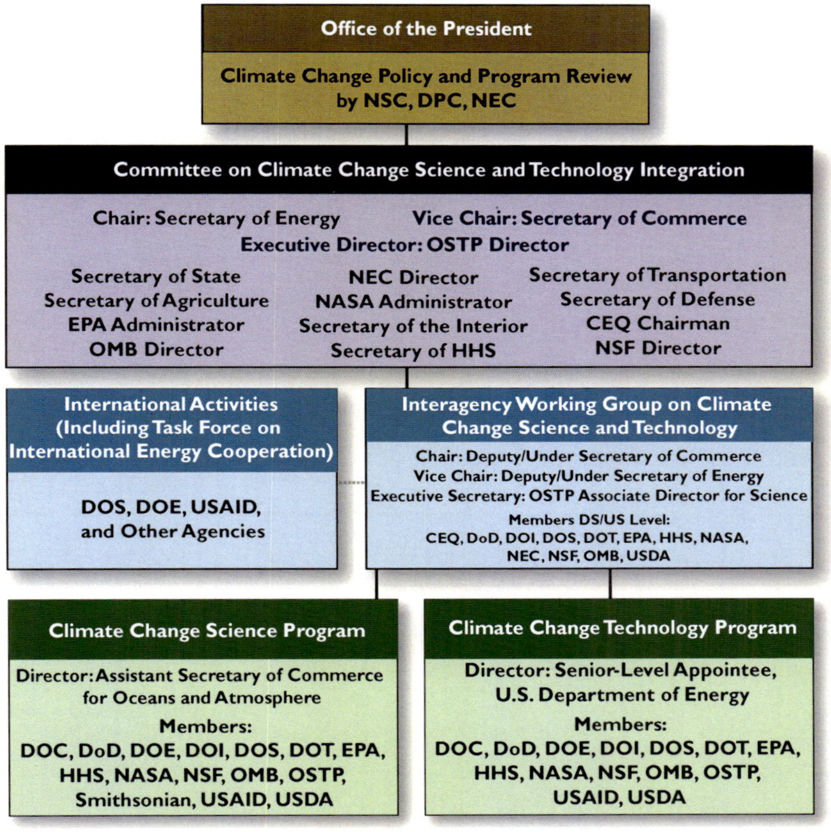

Note: NSC = National Security Council; DPC = Domestic Policy Council; NEC = National Economic Council; DS = Deputy Secretary; US = Under Secretary.

Reports of CCSP research appear as SAPs. Each has an identifying number related to the underlying element of the *Strategic Plan for the U.S. Climate Change Program.*[19]

The SAP reports, drawn from the deep expertise available in the United States, are documented, reviewed, and readily available on the USGCRP website.[20] Confidence in each finding follows a classification similar to that used by the IPCC, but it has not been applied as systematically. The strategic plan for the CCTP focuses on five technology areas:

- Energy efficiency
- Energy production
- Other gases

- Monitoring and measurement
- Sequestration.[21]

The CCTP is also based on the vision that most of the infrastructure that will emit GHGs in the coming century has not yet been constructed. This provides a significant opportunity to reduce emissions through strategic planning. We will discuss the impact of these technologies further in Chapter 4.

The GAO provides an additional significant source of critical analysis of the status of government programs with respect to climate change. Over the past decade, GAO has regularly pressed federal agencies to address climate change issues. Table 1-4 shows examples of reports and topics covered.

Table 1-4. GAO Reports Related to Climate Change

Report	Title	Programs addressed
GAO-08-265	*Expert Opinion on the Economics of Policy Options to Address Climate Change*	Cap-and-trade and carbon tax proposals before Congress
GAO-07-863	*Agencies Should Develop Guidance for Addressing the Effects on Federal Land and Water Resources*	Bureau of Land Management (BLM), U.S. Forest Service (USFS), U.S. Fish and Wildlife Service (USFWS), NOAA, and National Park Service (NPS)
GAO-07-285	*Financial Risks to Federal and Private Insurers in Coming Decades Are Potentially Significant*	National Flood Insurance Program, Federal Crop Insurance Corporation
GAO-07-1172	*Agencies Have Data-Sharing Policies but Could Do More to Enhance the Availability of Data from Federally Funded Research*	DOE, NASA, NOAA, NSF
GAO-06-97	*EPA and DOE Should Do More to Encourage Progress Under Two Voluntary Programs*	EPA Climate Leaders DOE Climate VISION

These and other GAO reports are essential reading to appreciate the recent history of climate change research in the United States and represent an independent assessment of agency efforts that have stimulated significant changes in management.

SUMMARY

Reliable international and national bodies conclude that the climate of the world is changing, humans are contributing to that change, and the change is already affecting important global systems, including food supply, transportation, energy production, and ecosystems. Temperature, precipitation, stream flow, snowpack, wind patterns, storminess, heat waves, and drought patterns in the United States and adjacent oceans are changing due to human influences. These changes affect natural and built environments that are the responsibility of the federal leader. The science supporting our understanding of that change, its impacts, and the options for mitigation and adaptation is convincing and reliable.

Six facts are key to understanding the human role in these changes: (1) GHGs in the atmosphere are now very rare but very effective in warming the earth; (2) human activity generates these GHGs and other GHGs never before present in the atmosphere; (3) the rarity of these gases means the human contribution greatly increases their amounts in the atmosphere; (4) as economies grow and produce more energy, the increase tends to accelerate each year; (5) these gases last a long time in the atmosphere; and (6) the abundance of these GHGs is changing the climate. Even if we significantly reduce our emissions of GHGs, the climate system will continue to change. Indeed, even if we stop our emissions of GHGs, we are already committed to significant climate change.

Few agencies currently demonstrate a commitment to addressing climate change through their strategic plans or budget documents. Policy options employed to mitigate climate change include (1) increasing energy efficiency, (2) fostering energy-efficient transportation, (3) encouraging conservation and renewable energy, and (4) sharing federal government expertise across all agencies.

Climate change already affects some agency operations, and these agencies must develop adaptation strategies. These include USDA (USFS), DOC (NOAA), DOT, and DOI (BLM, USFWS, and NPS). In the future, climate change will affect all agencies, and all will need to develop adaptation plans and implement those plans.

We base this guide on IPCC findings with high confidence (generally 90 percent or greater), and we supplement, clarify, and expand on the IPCC findings with related material from other sources (especially the CCSP and NAS) that is directly relevant to the concerns of the federal leader.

In Chapter 2, we review climate change science, as it specifically pertains to you as a federal leader. Chapter 3 compares national and regional programs that affect your agency. Later chapters illustrate capabilities you may want to develop in your organization, including carbon management (Chapter 4), impacts (Chapter 5), and adaptive measures (Chapter 6). We conclude this guide (Chapter 7) with a discussion of international issues in climate change to provide a wider perspective on these global-scale challenges.

---

[1] Intergovernmental Panel on Climate Change (IPCC), "Summary for Policymakers," in *Climate Change 2007: Synthesis Report. A Contribution of Working Groups I, II, and III to the Fourth Assessment Report of the Intergovernmental Panel on Climate Change*, Cambridge University Press, Cambridge, United Kingdom, and New York, NY, USA, www.ipcc.ch/pdf/assessment-report/ar4/syr/ar4_syr_spm.pdf.

[2] IPCC key finding. Throughout this book, each purple box shows a key finding from the IPCC with a high level of scientific understanding. See Appendix C for specific AR4 chapter sources for each finding. The beginning of this chapter gives a full discussion of our selection of key findings.

[3] IPCC, "Fact Sheet: How the Summary for Policymakers of an IPCC Working Group Report is Approved During the Plenary," *Information for the press*, www.ipcc.ch/press/ar4-factsheet2.htm.

[4] IPCC, "Procedures for the Preparation, Review, Acceptance, Adoption, Approval, and Publication of IPCC Report,." *Principles Governing IPCC Work*, Appendix A. Adopted at the Fifteenth Session (San Jose, April 15–18, 1999), amended at the Twentieth Session (Paris, February19–21, 2003) and Twenty-first Session (Vienna, November 3 and 6–7, 2003), www.ipcc.ch/pdf/ipcc-principles/ipcc-principles-appendix-a.pdf.

[5] To convert degrees Celsius to degrees Fahrenheit, multiply by 9/5, then add 32.

[6] Environmental Protection Agency (EPA), "Temperature Changes," *Climate Change—Science*, www.epa.gov/climatechange/science/recenttc.html, March 31, 2008.

[7] National Oceanic and Atmospheric Administration (NOAA), National Climate Data Center, *Global Surface Temperature Anomalies*, February 6, 2006, www.ncdc.noaa.gov/oa/climate/research/anomalies/anomalies.html.

[8] NOAA, *State of the Climate in 2007*, www.ncdc.noaa.gov/oa/climate/research/ 2007/ ann/bams/. See Chapter 2 for specifics on these and other observed changes.

[9] Climate Change Science Program (CCSP) and the Subcommittee on Global Change Research, *The First State of the Carbon Cycle Report (SOCCR): The North American Carbon Budget and Implications for the Global Carbon Cycle*, Synthesis and Assessment Product 2.2 [King, A.W. et al. (eds.)], NOAA, National Climatic Data Center, Asheville, NC, p. 22, Figure 2.1, www.climatescience.gov/Library/sap/sap2-2/final-report/default.htm.

[10] T. J. Blasing and K. Smith, *Recent Greenhouse Gas Concentrations*, Carbon Dioxide Information Analysis Center, September 2008, cdiac.ornl.gov/pns/current_ghg.html.

[11] See Chapter 2 for the changes in climate already observed and Chapter 5 for projected impacts.

[12] *Climate Change 2007: Synthesis Report. A Contribution of Working Groups I, II, and III to the Fourth Assessment Report of the Intergovernmental Panel on Climate Change,* p. 27, Cambridge University Press, Cambridge, United Kingdom, and New York, NY, USA, www.ipcc.ch/pdf/assessment-report/ar4/syr/ar4_syr_spm.pdf.

[13] Department of Energy (DOE), "Federal Energy Management Program," *Energy Efficiency and Renewable Energy*, www1.eere.energy.gov/femp/index.html.

[14] John B. Stephenson et al., *Climate Change: Agencies Should Develop Guidance for Addressing the Effects on Federal Land and Water Resources*, Government Accountability Office (GAO)-07-863 (Washington, DC: GAO, August 2007), www.gao.gov/new.items/d07863.pdf.

[15] EPA, "U.S. Climate Policy and Actions," *Climate Change—U.S. Climate Policy*, www.epa.gov/climatechange/policy/index.html, December 20, 2007.

[16] Executive Order (EO) 13423, *Strengthening Federal Environmental, Energy, and Transportation Management*, President George W. Bush, January 24, 2007, www.whitehouse.gov/news/releases/2007/01/20070124-2.html.

[17] EO 13432, *Cooperation Among Agencies in Protecting the Environment with Respect to Greenhouse Gas Emissions From Motor Vehicles, Nonroad Vehicles, and Nonroad Engines*, President George W. Bush, May 14, 2007, www.whitehouse.gov/news/releases/2007/05/20070514-1.html.

[18] U.S. Global Change Research Program (USGCRP), *About the U.S. Global Change Research Program*, www.usgcrp.gov/usgcrp/about/default.htm, June 22, 2007.

[19] USGCRP, *Strategic Plan for the U.S. Climate Change Program*, July 2003 (update May 2008).

[20] USGCRP, *Announcements from the U.S. Climate Change Science Program*, November 12, 2008, www.usgcrp.gov/usgcrp/default.php.

[21] U.S. Climate Change Technology Program, *Strategic Plan*, DOE/PI-0005, September 2006.

Chapter 2
The Science of Climate Change

Understanding the carbon cycle and the behavior of the climate system is a precursor to mitigation and adaptation—the two possible actions to address the challenges of climate change. Such knowledge comes from the answers to five questions:

1. Are we changing the carbon cycle?
2. Does changing the carbon cycle change the climate?
3. How does the climate system respond?
4. Is the climate changing now?
5. How will the climate change in the future?

> Continued GHG emissions at or above current rates would cause further warming and induce many changes in the global climate system during the 21st century that would very likely be larger than those observed during the 20th century.[1]

ARE WE CHANGING THE CARBON CYCLE?

What Is the Carbon Cycle?

When carbon combines with oxygen to form CO_2, it is found in the atmosphere and soils, and dissolved in water bodies. Carbon, in other forms, is also a component of the fossil fuels we burn to release energy. CH_4 contains carbon, and both natural regions (soils, wetlands, and sedimentary deposits) and human-perturbed ones (farmlands, feedlots, landfills, and fuel infrastructure) generate CH_4.

> It is very likely that the observed increase in CH_4 concentration is predominantly due to agriculture and fossil fuel use.[1]

Figure 2-1 illustrates the cycling of carbon among various reservoirs. Carbon is either stored or released, depending on a variety of natural processes as well as human influences. Carbon reservoirs include the following:

- *Atmosphere*. Gases such as CO_2 and CH_4.
- *Biosphere*. Organic matter, plants, and soils.

- *Hydrosphere*. Streams, rivers, lakes, and oceans.
- *Lithosphere*. Fossil fuels, sediments, and rocks.

Figure 2-1. Carbon Fluxes and Reservoirs in the Global Carbon Cycle

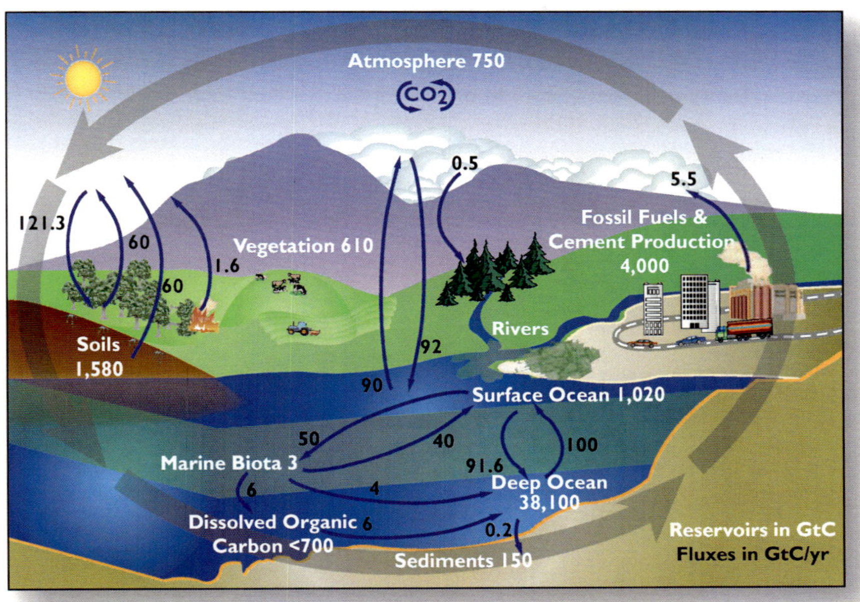

Note: The numbers in white represent reservoirs (in Gt of carbon). The numbers in black list the fluxes of carbon between various parts of the global system (in Gt of carbon per year).

Carbon is exchanged between all reservoirs, often in both directions. When the amount of carbon stored in one of these reservoirs changes, it becomes a source or sink for carbon elsewhere. A carbon *sink* absorbs more carbon than it releases. Conversely, a carbon *source* releases more than it absorbs.

For example, if a forest burns it has been changed from a sink to a source. This shifts the carbon balance and requires other sinks to absorb the excess carbon, resulting in fluxes between the reservoirs (the black numbers in Figure 2-1).

The largest reservoir of carbon is the deep ocean and the next largest is fossil fuels. Compared with those reservoirs, the atmosphere stores only a small amount of carbon. Thus, a small fraction shift to the atmosphere from one of these reservoirs can greatly increase the atmosphere sink.

Carbon moves more slowly through some reservoirs than others. Generally carbon in the deep ocean has a long residence time. In soils, some carbon cycles quickly, but some humus in soils holds carbon a long time. Carbonate rocks such as limestone store vast quantities of carbon in a stable form that cycles very slowly. The residence time of CO_2 in the atmosphere is 150–500 years.

The residence time is the size of the reservoir divided by the flux, both shown in Figure 2-1.

Similar to most budget processes, the global carbon budget is an accounting system that tracks the fluxes (both positive and negative) between reservoirs. In a stable cycle, the input of carbon to each reservoir equals the output. However, for more than 100 years, the flux to the atmosphere has exceeded the flux out of the atmosphere, increasing the atmospheric concentration of CO_2 (and CH_4).

The carbon budget is now out of balance due to the accumulation of carbon in the atmosphere in the form of CO_2 and CH_4, primarily from the use of coal, petroleum, and natural gas and agriculture and land use changes.

How Has the Carbon Cycle Changed?

About half of the excess flux has remained in the atmosphere. Some has moved to the surface waters of oceans (about 30 percent), and about 20 percent has moved to forests and soils. This readjustment of the carbon cycle leads to major changes in biogeochemical processes in these other reservoirs. Because of fossil fuel emissions, North America is currently a net source of CO_2 to the atmosphere, contributing to the global buildup of CO_2.[2]

If you sum the reservoirs and fluxes in Figure 2-1, you will find that they are not balanced. Sometimes when you try to balance your checkbook, you notice that your calculations are off and therefore try to find the missing funds. In the same sense, when scientists try to balance the carbon budget, they find that the atmosphere retains less of the CO_2 emissions than can be explained by the known fluxes to other reservoirs from the atmosphere. This CO_2 must flux to another reservoir, but we don't observe that flux and don't know which reservoir receives the excess from the atmosphere or if there is another sink that has yet to be identified. We have a good idea where the carbon is going in general, but not why and where in specific.

Figure 2-2 shows North American carbon sources and sinks (in Gt of carbon per year) in 2003. Error bars show the 95 percent confidence interval for each estimate. As can be seen, the input from the use of fossil fuels dwarfs that of the natural elements. We think that most natural elements are sinks for the excess carbon being added to the atmosphere, but we are uncertain. Forests are the major sink, but their uptake of carbon may change in the future due to biochemical limitations and adaptations.

Figure 2-2. North American Carbon Sources and Sinks

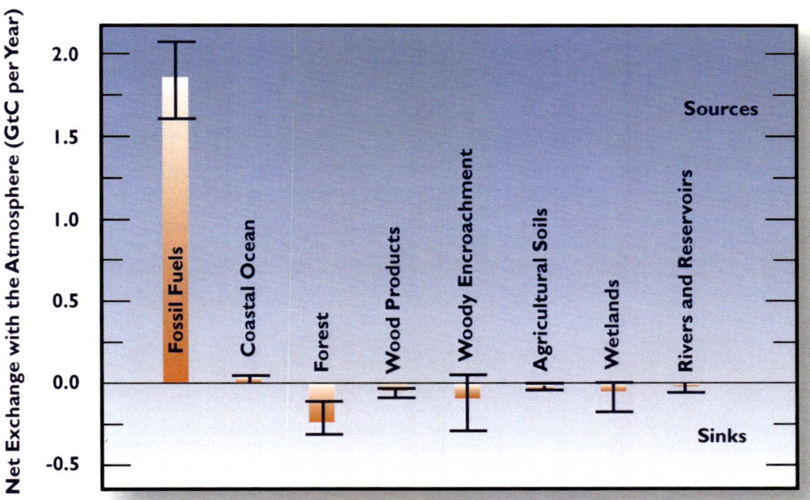

Will the Carbon Cycle Continue Changing?

The carbon cycle balance also has a dynamic aspect. Various elements of the climate system act at different rates (see "How Does the Climate System Respond?"). For example, the movement of CO_2 from atmosphere to ocean (which occurs by chemical, physical, and biological actions at the ocean surface) may be slow compared with the size of that sink. It will take thousands of years for the oceans to reach equilibrium with the CO_2 now in the atmosphere or that humans may contribute in the coming century.

The transfer of CO_2 from the surface ocean, which can hold only a small amount of CO_2, to the deep ocean, which can hold an enormous quantity of CO_2, is a bottleneck that will delay equilibration of the climate system for many generations. This transfer depends on complex circulation patterns that themselves could be disrupted. Meanwhile, the ocean will also be equilibrating to the transfer of more heat from the atmosphere, another slow process. And those two processes interact because warm water holds less CO_2.

As the oceans absorb CO_2 from the atmosphere, the CO_2 dissolves and the concentration of hydrogen ions in seawater increases, thereby decreasing pH (increasing the acidity) of the ocean. This process is referred to as ocean acidification, which in turn affects marine species (see Chapter 5). So even though the oceans function as a major carbon sink for the increased human-caused CO_2 emissions and help to mitigate climatic effects, there are additional biological, physical, and chemical impacts on the ocean.

Because CO_2 is essential for photosynthesis, elevated levels of CO_2 can "fertilize" plants and increase their productivity. However, most plants are limited not by CO_2 but by the availability of nitrogen (N_2). So the response of a plant to CO_2 also depends on the availability of N_2. The CO_2-enhanced growth increases the use of N_2, eventually reducing soil nitrogen availability which then limits plant growth response to CO_2. Rising CO_2 will very likely increase photosynthesis in young forests on fertile soil.[3] Internal combustion engines, some industrial processes, and application of fertilizers all increase the amount of N_2 available.

Plants receiving human contributions of N_2 may experience growth while others may not. Much of this nitrogen washes into the ocean and increases microbe metabolism at depths, which depletes waters of oxygen and creates "dead zones" and changes CO_2 uptake.

Uptake of CO_2 by forests and by soil organisms is an important sink of CO_2 in North America. Because of the complex biogeochemistry involved, we don't know how long this will continue and how much excess CO_2 this sink can absorb. When it reaches capacity, other sinks (such as the atmosphere) will accumulate faster, which means, *even with constant emissions*, the rate of increase of CO_2 in the atmosphere will ultimately rise.

The rate of increase of CO_2 in the atmosphere will rise even if emissions are kept constant once certain sinks reach their storage capacity.

Large stores of carbon are tied up in bogs and wetlands as well as frozen ground in high northern latitudes. Rising temperatures release some of this carbon. It could remain a slow but significant carbon sink, or reverse and become a source, rapidly releasing large amounts of CO_2 and CH_4 to the atmosphere.[2]

The disturbance of the carbon cycle, the resulting buildup of atmospheric gases, and the complex chains of impacts and feedbacks are the major driver of climate change.[2] All elements of the climate system interact. As human-caused CO_2 emissions increase, effects are observed throughout the climate system and are very likely to increase in the years to come. This awesomely complex system behavior can only be understood with a rigorous and systematic approach.

> In order to stabilize the concentration of GHGs in the atmosphere, emissions would need to peak and decline thereafter. The lower the stabilization level, the more quickly this peak and decline would need to occur. There is high agreement and much evidence that mitigation efforts over the next 2 to 3 decades will have a large impact on opportunities to achieve lower stabilization levels.[1]

Does Changing the Carbon Cycle Change the Climate?

The recent cause for concern regarding climate change is the rapid increase in global surface temperature over just a few decades with no known natural explanation. Evidence suggests that human activity is playing a leading role in the increase of global surface temperature, which greatly impacts the natural processes and interactions of the climate system.

Energy Balance: Driver of Climate Change

Increasing CO_2 in the atmosphere changes the energy balance of the earth. Figure 2-3 illustrates how the earth balances energy radiated from the sun. Incoming solar radiation is absorbed by the atmosphere, absorbed by the surface, or reflected back to space. Over time, the amount of incoming solar radiation (short wave) balances energy emitted (long wave/infrared) from the earth.

The numbers in Figure 2-3 (expressed in watts per square meter, or W/m^2) represent the amount of energy transferred by each radiation process. Short-wave radiation from the sun passes through the atmosphere and is absorbed by the earth, warming it. That warmth is expressed as long-wave energy and is reradiated to the atmosphere.

> *We can convert from W/m^2 to sun hours per day by multiplying by 0.024, so 1.6 W/m^2 is equivalent to about 2.3 extra minutes of sun each day. This adds up to cause significant warming, just as a handful of extra calories in your daily diet would eventually cause significant weight gain.*

GHGs present in the atmosphere both absorb and emit long-wave radiation (infrared) at a variety of wavelengths, defining the energy balance of the earth. Radiation absorbed by GHG molecules elevates them to a more energetic state. They return to a less excited state by emitting infrared radiation—into space and toward the surface of the earth. This process (known as the greenhouse effect) maintains the temperature of the earth at a level that sustains life. A greater concentration of GHG in the atmosphere absorbs and re-emits more heat to the earth so global temperature increases.

Figure 2-3. Energy Balance of Earth (in W/m²)

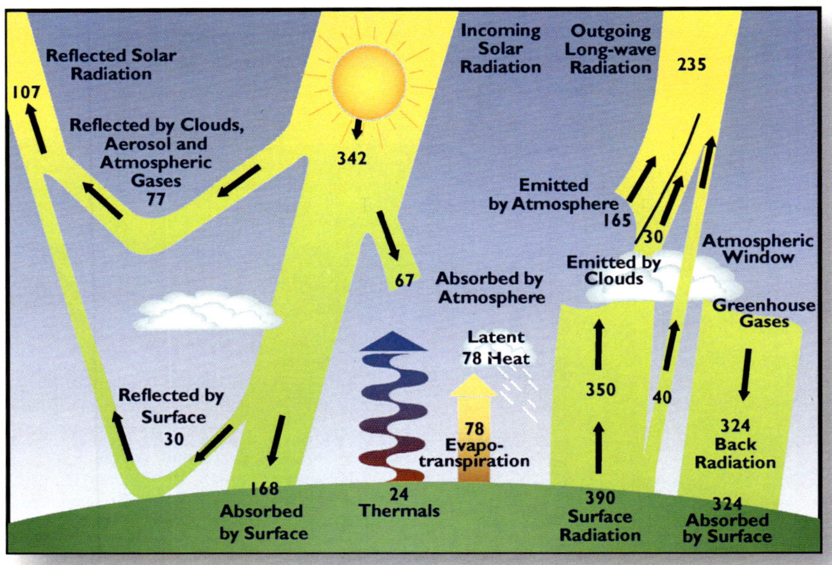

The imbalance of incoming solar radiation and outgoing infrared radiation creates a *radiative forcing*. The term forcing conveys the idea that the earth's radiative balance changes from its previous state by an intervention. A positive forcing warms the earth and a negative forcing has a cooling effect. Both positive and negative forcings occur.

Increasing concentrations of GHGs are the most significant known positive radiative forcings now at work in our atmosphere, and most of this increase comes from the burning of fossil fuels and deforestation.

> The combined radiative forcing due to increases in CO_2, CH_4, and N_2O is +2.3 [+2.1 to +2.5] W/m^2, and its rate of increase during the industrial era is very likely to have been unprecedented in more than 10,000 years. The CO_2 radiative forcing increased by 20 percent from 1995 to 2005, the largest change for any decade in at least the last 200 years.[1]

What does an increase in average global surface temperature of a few degrees Celsius really mean?

Although an increase of a few (say 2–3) degrees Celsius does not sound like much cause for concern, if you look at the annual average temperatures of certain U.S. cities, it would be like trading the average climate of Washington, DC, for the average temperature of Atlanta, GA. Similarly, if you lived in Juneau, AK, you would be experiencing a temperature similar to that of Portland, ME.

To put it in a different perspective, according to IPCC, the change in global average temperature between the last glacial and interglacial period was 4–7˚C over a period of 5,000 years. Therefore, if an increase of 4˚C can melt ice sheets covering large parts of the earth, the projected increase of a few degrees Celsius over only a handful of decades is *significant*.

The uncertainty in the radiative forcing [in brackets in the previous purple box] is due mostly to uncertainty about the effects of aerosols. Anthropogenic contributions of aerosols in the atmosphere produce cooling effects. However, these cooling effects are about half of the warming effects due to human-produced GHGs, causing a net warming.

When the values of both positive and negative radiative forcing are taken together and all interactions among climatic factors are accounted for, *the total net increase in surface energy due to human activities is 1.6 W/m^2*. This small amount is cumulative and leads to the vast changes taking place in our climate.

There is very high confidence that the global average net effect of human activities since 1750 has been one of warming, with a radiative forcing of +1.6 [+0.6 to +2.4] W/m^2.[1]

The temperature response to this forcing is complicated by the presence of ozone (O_3). O_3 in the stratosphere (6–30 miles up, the second layer of the atmosphere, just above the troposphere) absorbs UV radiation. This shielding layer of O_3 has been gradually depleted by human-made chemicals like CFCs.[4]

Since O_3 absorbs sunlight and warms the stratosphere, its depletion can have a cooling effect. This is why, as the earth's surface temperatures reach record highs, record lows are taking place in the stratosphere. This produces a negative radiative forcing but not enough to balance the positive forcing of increased CO_2.

The observed pattern of tropospheric warming and stratospheric cooling is very likely due to the combined influences of GHG increases and stratospheric ozone depletion.[1]

A Critical Experiment

There are two means of comparing the evolution of climate change—through the observed record and through models, which illustrate how climate could react to changes over time on the basis of certain scenarios. Climate is always changing, so who is to say that humans have an influence? Scientists sought to answer this question by designing a critical experiment.

Imagine (hypothesize) that humans have no effect on climate. If that is true, then we can represent the temperature changes of the past 100 years by considering only natural processes. So scientists used their most sophisticated computer model of the natural climate system to calculate how climate should have varied over the past 100 years, taking into account all the known variations in natural forcings such as volcanic eruptions.

To represent the uncertainty in their understanding of the climate system, they reran the model many times, each time slightly changing the parameters but always keeping them in the range of the natural system (an ensemble). They could not reproduce the observed pattern of temperature with any of these model results—as shown where the blue line (observations) diverges around 1960 in Figure 2-4(a).[5] Maybe this just means our models are bad?

What if we can match the record if we include the known human radiative forcings? This would suggest the lack of fit in Figure 2-4a is not due to a bad model but due to failure to take human forcings into account. So the modelers repeated the experiment but this time included all the known anthropogenic radiative forcings. Again, they repeated the experiment many times, each time slightly varying the parameters to represent their uncertainty in the various model parameters but always keeping them within the range that represents human forcings.

This time the model results clearly bracket the observed pattern of temperature change (Figure 2-4b), so it appears we can only understand the record of global temperature change during the past 100 years by assuming that anthropogenic forcing in the form of GHGs is at work. This elegant experiment is a critical result and strongly supports the accumulating body of evidence that humans are affecting the earth's climate (particularly over the past 50 years).

Figure 2-4. Results of Global Temperature Change Experiments

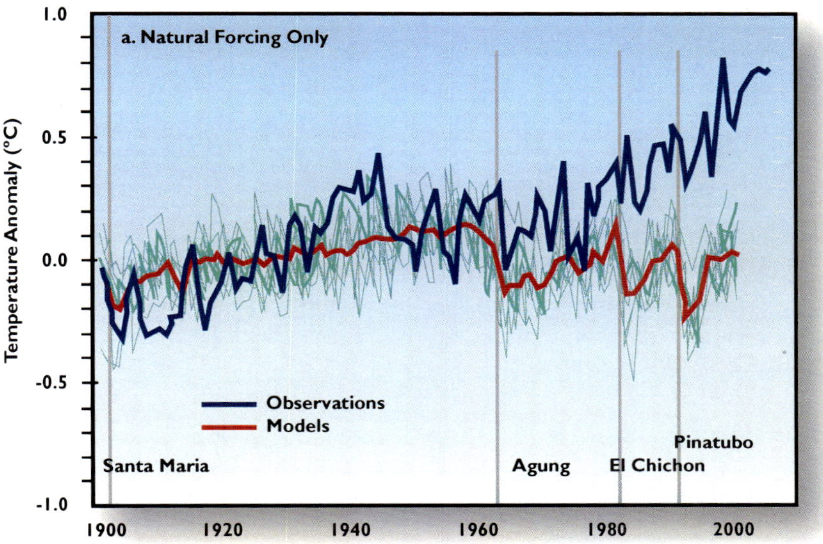

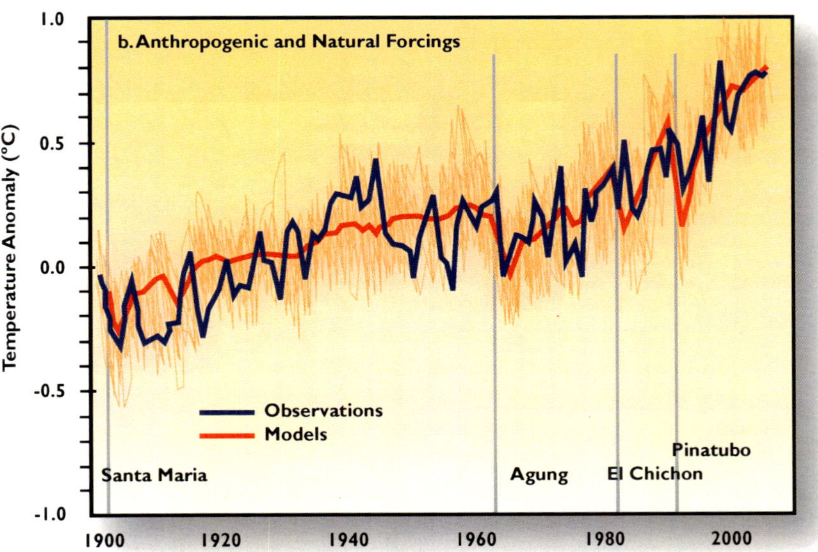

Note: Vertical lines indicate the volcanic eruptions of Santa Maria (1902), Mount Agung (1963-64), El Chichon (1982), and Mount Pinatubo (1991).

> The observed widespread warming of the atmosphere and ocean, together with ice mass loss, support the conclusion that it is extremely unlikely that global climate change of the past 50 years can be explained without external forcing, and very likely that it is not due to known natural causes alone.[1]

HOW DOES THE CLIMATE SYSTEM RESPOND?

Climate continuously changes: this is a natural byproduct of the feedbacks, step changes, amplifiers, and time lags of a complex system. The glacial and interglacial periods that the earth has experienced over nearly 2 million years are evidence that our climate is subject to natural extreme changes over time; however, the rate at which that change occurs is of the magnitude of thousands of years or more (see Figure 1-3).

Various parts of the climate system (Figure 2-5) change at different rates. The unstable atmosphere changes fastest and over the greatest range. In that case, changes can include temperature, moisture content, wind speed, and pressure. The biosphere responds in many ways (growth, decomposition, N_2 fixation, transpiration, etc.) and at many rates, depending on species and location. Oceans change more slowly, ice bodies slower still, while the solid earth is most stable.

These different rates of change produce lags in response of system components. Buildup of GHGs in the atmosphere only slowly changes the response of ecosystems and the composition of the ocean, and a slowly warming ocean instills ponderous changes in the large and stable ice sheets. But when ice sheets melt, slow motions in the mantle may lead to the rebound of the earth's crust that was depressed by the weight of that ice. The original buildup of GHGs can occur in decades, the resultant rebound may take thousands of years. Because of these lags, climate is constantly changing; it's always trying to catch up to what happened before!

These lags contribute to the complexity of the system and imply that the system today is already responding to past changes, things initiated in the natural system well before humans began to impact the atmosphere. The earth is rebounding in some places (Great Lakes, Maine, and Alaska) from previous glacial ages. The ice sheets are still responding to climates of the past even as new changes complicate that response. Ocean circulation reflects ice sheet and ocean basin configurations of the past.

Figure 2-5. Climate System Components, Processes, and Interactions

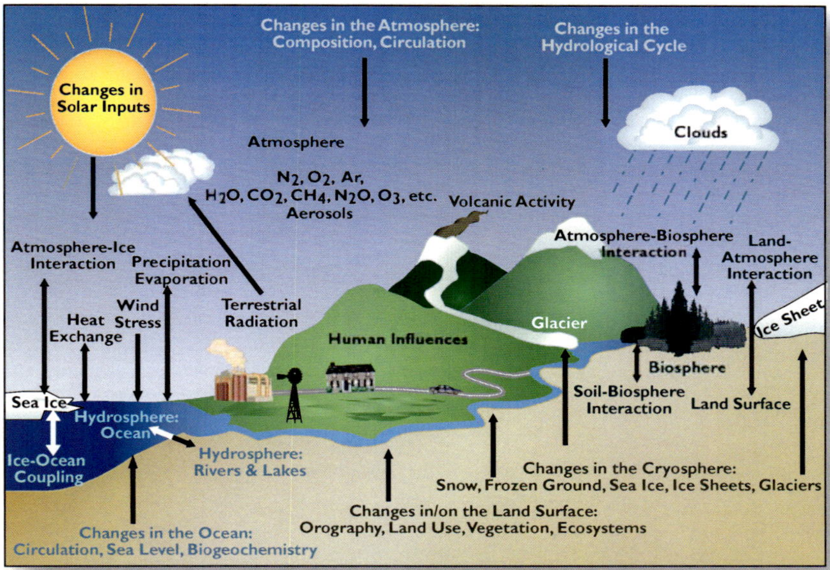

The 2,000-year-old redwood tree began life when oceans, ice sheets, and earth's rebound were in very different states, yet it must photosynthesize under CO_2 concentrations greatly increased by human actions today. Such lags in response are the norm in natural systems and mean changes we make today (such as an increase in CO_2 emissions) will affect the system far into the future.

Changes we make today will affect the system far into the future.

Small responds quickly, large responds more slowly. It is faster to heat a tea cup than a tub. Smaller bodies of water warm faster than large ones, lakes warm faster than oceans, and glaciers warm faster than ice sheets. This is one reason why different parts of the same system change at different rates. The South Pacific Ocean is far more stable than the North Atlantic Ocean, which is far more stable than the Great Lakes, which are far more stable than the pond in your neighborhood park, which also changes faster than a mountain glacier of the same size (ice is more stable than water).

The vast number of parts of the climate system of different sizes and response times contribute variations at many time scales and with many different lag times in the system. No wonder, then, that a record of climate variability will have a complex and erratic appearance. The complexity of the system means that some parts will be more vulnerable to change and that the timing and magnitude of changes will differ from one part to another.

Atmosphere

The unstable atmosphere consists of components that respond to solar radiation at different rates and intensities. N_2, oxygen (O_2), and argon (Ar) constitute the greatest part. O_2 is a strong absorber of incoming solar radiation, while N_2 is chemically reactive and Ar is inert. Water vapor, CO_2, CH_4, and N_2O, known collectively as GHGs, are naturally present in trace amounts in the atmosphere, except for water, which is more abundant but highly variable. There are additional human-manufactured GHGs (see Table 1-1).

GHGs absorb heat energy from the earth that otherwise would exit to space. When they reradiate it, about half returns to earth, leading to higher surface temperatures.

Important chemical reactions in the atmosphere influence GHG abundance. For example, CH_4 is oxidized to CO_2. Figure 2-1 illustrates the flow of CO_2 and CH_4 between the atmosphere, hydrosphere (water bodies), cryosphere (ice sheets and glaciers), land surface (sediment and soils), and biosphere (marine and terrestrial). Much of this flow is mediated by biogeochemical reactions.

Hydrosphere

The hydrosphere includes all water bodies, both fresh and salt water. Water runoff from land alters the volume, composition, and temperature of water bodies. Water absorbs gases such as CO_2, which makes it acidic. Warm water holds less CO_2 than cold water, a relationship that is unusual for gases, and means that rising temperatures will mean more CO_2 released to the atmosphere.

Warming water evaporates faster; cooling water vapor leads to condensation and precipitation. So energy input drives the hydrologic cycle (condensation → precipitation → runoff → streamflow → ocean circulation → evaporation → condensation). Greater energy input (higher temperatures) accelerates the hydrologic cycle.

Warmer air holds more moisture, so it can take longer to reach the level at which precipitation begins, and intervals between storms (dry periods and droughts) can increase. And when this enhanced moisture is released, it can produce more intense storms. A positive feedback arises because moist air holds more heat.

Warmer global temperatures raise sea level by expanding ocean water, melting mountain glaciers and small ice caps, and causing portions of the coastal section of the Greenland and Antarctic ice sheets to melt or slide into the ocean.[6] Sea level can also rise or fall due to local tectonic changes in the Earth's crust. These tectonic changes can include rebound as ice sheets melt.

> New data since the IPCC Third Assessment Report now show that losses from the ice sheets of Greenland and Antarctica have very likely contributed to sea level rise from 1993 to 2003.[1]

Ocean Circulation

Just as heated air expands and rises, large water bodies develop circulation patterns driven by differences in temperature and density. Warmer water is less dense and rises. More salt makes ocean water denser and it sinks, so oceans move in response to salinity and temperature (thermohaline circulation).

Greater evaporation makes the ocean saltier, greater precipitation makes the ocean fresher. So increasing energy input (higher temperature) increases density differences and accelerates thermohaline circulation.

Figure 2-6 illustrates the great ocean "conveyor belt" that transports heat across the globe. The part of this conveyor belt receiving the most attention due to potential climatic changes is in the North Atlantic. Warm water from the tropics is transported via this conveyor belt to the waters of the North Atlantic via the Gulf Stream, where heat is slowly released. As the water moves toward western Europe, the surface water evaporates and the ocean releases heat to the atmosphere (hence, the relatively mild climate in western Europe).

In the higher latitudes, the ocean loses heat. As the water evaporates, the salinity increases, causing the water to become denser. The water then sinks deep below the surface leading to a large-scale deep overturning motion of the ocean, referred to as meridional overturning circulation (MOC). This water is then recirculated through the conveyor belt, emerging again thousands of years later.

Figure 2-6. Thermohaline Circulation of Global Ocean System

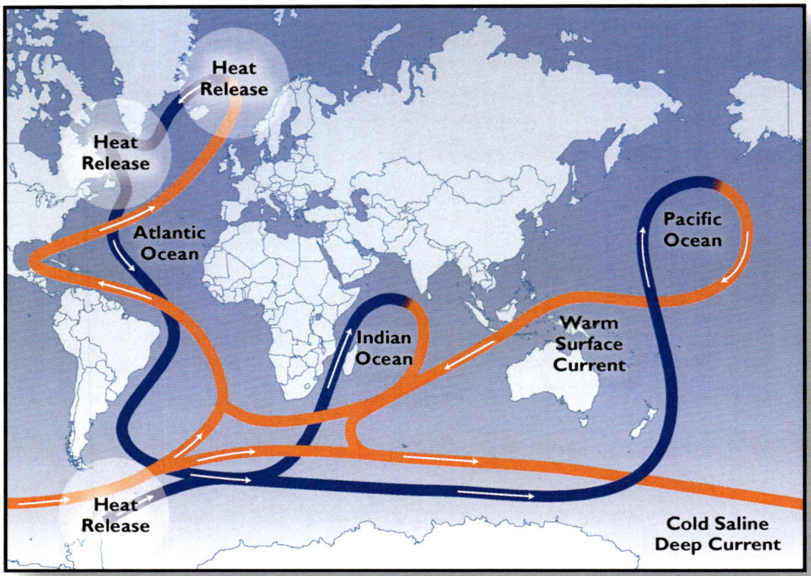

Slowdown of the conveyor belt would cool western Europe (due to slower heat transport in the conveyor belt) despite the increase in global temperature.

> Based on current model simulations, the MOC of the Atlantic Ocean will *very likely* slow down during the 21st century; nevertheless temperatures over the Atlantic and Europe are projected to increase. The MOC is *very unlikely* to undergo a large abrupt transition during the 21st century.[1]

Cryosphere

The cryosphere comprises water in solid form such as glaciers, ice sheets, snow cover, permafrost, freshwater ice, and sea ice. These ice bodies contain significant amounts of water evaporated from the ocean. When that water precipitates as snow and is added to ice sheets, sea level is lowered, and when the ice melts, sea level rises.

The accelerated hydrologic cycle and higher temperatures discussed above will change the dynamics of the cryosphere. Some ice bodies will shrink and disappear, others may grow.

Ice has high albedo (ability to reflect solar radiation). Land with low albedo, exposed when snow or ice melts, absorbs more solar radiation and surface temperature rises: higher temperature → more melting → less ice → less reflection of sunlight → more energy absorbed → higher temperatures → more melting. This cycle represents a fundamentally important positive feedback in the climate system and one that appears to be at work at high northern latitudes today.

This same positive feedback occurs when sea ice (ice floating on water as in the Arctic Ocean) melts. Positive feedback makes systems less stable. High northern latitudes, which feature much ice, are warming the most today. We see changes in southern polar latitudes also, particularly in the extent and duration of sea ice.

The three major ice sheets (Greenland, West Antarctic, and East Antarctic) each respond differently to climate change. Greenland is largely bordered by high mountains and its ice only drains through narrow passageways in ice streams directly into deep ocean water.

The West Antarctic ice sheet is relatively low in elevation and has a relatively warm base that lubricates its flow. Its draining ice streams are buttressed by floating ice and could easily become destabilized. Meltwater from this ice sheet can affect deep-water circulation patterns.

The thick East Antarctic ice sheet holds the most ice and its surface elevation is so high, temperatures are well below freezing. Even warming of 4°C would not lead to significant loss of mass of that ice sheet.

Biosphere

CO_2 is a byproduct of biological processes, past and present. Plants taking carbon from CO_2 have formed all of our carbon fuels (coal, oil, natural gas, and wood); burning those fuels releases CO_2. Many living organisms on land and in surface waters of the oceans fix carbon, and most depend on the energy stored in carbon bonds for energy.

Because of this dependence of life upon carbon bonds, carbon is continually cycling through the ecosystems of the earth, and a dynamic balance exists between the amounts and fluxes of carbon in various parts of the biosphere. That balance is disrupted when excess CO_2 is available. The imbalances are magnified when other elements critical to carbon fixation (such as nitrogen) are artificially injected into ecosystems. Human activity has doubled the available N_2 in natural systems.

The marine and terrestrial biospheres each play a major role in the composition of the atmosphere. Through photosynthesis, plants and microorganisms on land and in the ocean store large amounts of carbon drawn from CO_2 in the atmosphere. Deforestation is now the second largest human contribution of CO_2 to the atmosphere after fossil fuel burning.

The marine biosphere is especially prolific in coastal areas most affected by human activity. Productivity in the open ocean is controlled by light, water temperature, and the availability of nutrients, which are generally scarce far from riverine sources. N_2, phosphorus, and iron are the primary limiting nutrients to carbon fixation in the open ocean.

The carbon cycle interacts intimately with other biogeochemical cycles (especially those of N_2, oxygen, phosphorus, and sulfur), each of which is disrupted by human activities.[2]

Vegetation and soil control how surfaces absorb, reflect, and reradiate solar energy. Changing land use or land cover affects both solar absorption and the partition between reflection and reradiation. This in turn changes surface temperature and thus the amount of moisture evaporated from soils and transpired by plant leaves. An increase in soil moisture content affects surface temperature through evaporation, which transports heat to the atmosphere.

The texture or roughness of the land, which varies with vegetative cover, influences friction and changes wind speed, turbulence, and transportation of particulates to the atmosphere.

Is the Climate Changing Now?

As explained in Chapter 1, warming of the climate system is “unequivocal” according to the IPCC. Observations of increases in global average air and ocean temperatures, widespread melting of snow and ice, and rising global average sea level now make this evident. The changes that come in response to this overall warming are complex but real, and the impacts are being documented around the world.

What creates the biggest concern about these changes is that they tend to be consistent with the picture expected if the known increase in GHGs is the major driver.

Temperatures

Global surface temperatures have increased about 0.74°C ±0.2°C since the late-19th century, and the linear trend for the past 50 years of an increase in 0.13°C per decade is nearly twice that for the past 100 years.

The 1990s were the warmest decade in the observational record. Seven of the 8 warmest years on record have occurred since 2001 and the 10 warmest years have all occurred since 1995.[7]

The term global warming can be misleading at times because it implies that one should assume all regions in the world experience increases in surface temperature, whereas some regions actually experience surface cooling.

The warming has not been globally uniform. Some areas (including parts of the southeastern United States and parts of the North Atlantic) have actually cooled slightly over the last century. The recent warmth has been greatest over North America and Eurasia between 40°N and 70°N.[7]

The ocean has absorbed 80 percent of the added heat, with average ocean temperatures increasing to depths of at least 3,000 meters.[8]

Average Northern Hemisphere temperatures during the second half of the 20th century were very likely higher than during any other 50-year period in the last 500 years and likely the highest in at least the past 1,300 years.[1]

Widespread changes in extreme temperatures have been observed over the last 50 years. Cold days, cold nights, and frost have become less frequent, while hot days, hot nights, and heat waves have become more frequent since 1950.[8]

The number of precipitation events that lead to flooding has increased, but not in every region. The extent of regions affected by droughts has also increased as precipitation over land has marginally decreased while evaporation has increased due to warmer conditions.[8]

Ice and Frozen Ground

The annual average Arctic sea ice extent shrank by 2.7 percent per decade, and the coverage in summer decreased 7.4 percent during the past several decades. The large recent loss of thick multiyear ice implies that this warming effect at high latitudes will persist.

Temperatures at the top of the permafrost layer have generally increased since the 1980s by up to 3°C, and the maximum area covered by seasonally frozen ground has decreased by about 7 percent in the Northern Hemisphere since 1900. In spring, the area covered by frozen ground decreased up to 15 percent.

In the Western United States, decreasing mountain snowpack and earlier snowmelt have altered the timing of stream flows.[9] This has significant implications for flood control, irrigation, and summer drying of vegetation.

Sea Level

Human influences have very likely contributed to sea level rise during the latter half of the 20th century.[1]

Losses from the Greenland ice sheet accelerated in 2003–05.[10] With the loss of buttressing sea ice, glacial flows have sped up in the past decade in parts of western Antarctica.[11] This meltwater raised sea level. With increasing global average temperature and the resultant increase in ocean temperature, ocean waters expanded and sea level rose. Table 2-1 summarizes the relative contributions to sea level rise from these various sources.

Table 2-1. Observed Rate of Sea Level Rise and Estimated Contributions

Source	Rate (mm/year)	
	1961–2003	1993–2003
Thermal expansion	0.42 ±0.12	1.6 ±0.5
Glaciers and ice caps	0.50 ±0.18	0.77 ±0.22
Greenland ice sheet	0.05 ±0.12	0.21 ±0.07
Antarctic ice sheet	0.14 ±0.41	0.21 ±0.35
Sum of individual climate contributions	1.1 ±0.5	2.8 ±0.7
Observed total sea level rise	1.8 ±0.5	3.1 ±0.7
Difference	0.7 ±0.7	0.3 ±1.0

About 150 million people live within 1 meter of sea level.

Figure 2-7 shows how the sea level has risen 100 to 200 mm[12] during the period when instrumental records (tidal gauges) have been available (shown in red with the uncertainty band in light red). A reconstructed range of sea level based on preserved records is shown for the period before 1870 when the first reliable records became available. They suggest that sea level was relatively stable in that period.

In agreement with climate models, satellite data and hydrographic observations show that sea level is not rising uniformly around the world. In some regions, rates are up to several times the global mean rise, while in other regions sea level is falling.

Figure 2-7. Past Global Average Sea Level

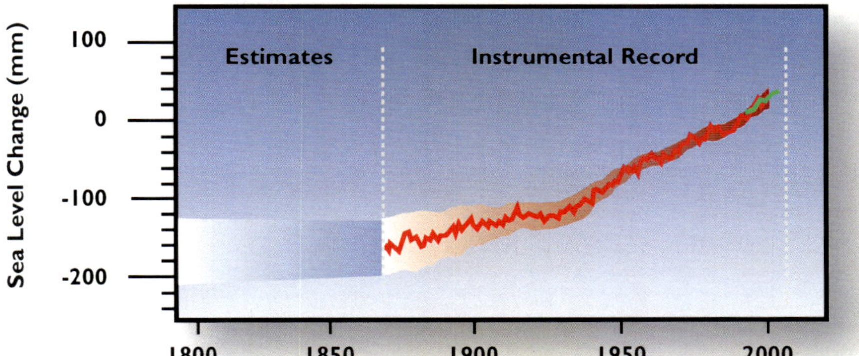

Biological

Altered timing of spring events of plants and animals has been reported for a broad range of species and locations, primarily in North America, Eurasia, and Australia. IPCC documented that leaves have unfolded earlier in North America and Europe in the last 30 years. Return dates of migrant birds have also shown changes in recent decades associated with changes in temperature in wintering or breeding grounds or changes in the migration route, and egg-laying dates have advanced.

Temperature increases have strongly influenced the distribution and abundance patterns of both plants and animals, due to the physiological constraints of species. Some plant and animal species are being observed at higher latitudes than previously reported.

The IPCC's third assessment report found effects of climate change on some species, but AR4, released only 6 years later, found an even wider range of species was being affected around the world than previously thought.

> There is very high confidence, based on more evidence from a wider range of species, that recent warming is strongly affecting terrestrial biological systems, including such changes as: earlier timing of spring events, such as leaf-unfolding, bird migration, and egg-laying; poleward and upward shifts in ranges in plant and animal species.[1]

Species and ecosystems are directly affected by changes in climate. Marine and terrestrial habitats have been affected by increasing surface and ocean temperatures, recession of glaciers, increased frequency and intensity in drought, and more intense storms, as well as ice backup in rivers. Generally speaking, in regions where a warming trend has been observed over the past few decades, changes in species have been observed.

> A synthesis of studies strongly demonstrates that the spatial agreement between regions of significant warming across the globe and the locations of significant observed changes in many natural systems consistent with warming is very unlikely to be due solely to natural variability of temperatures or natural variability of the systems.[1]

Living organisms, oceans, temperature patterns, and sea levels have changed due to the impact of increased concentrations of GHGs in the atmosphere from human emissions. Will these changes continue? Looking into the future requires the use of models; in the next section, we see how those models are used and what they tell us.

How Will the Climate Change in the Future?

The scientific and policy communities rely on climate models to provide detailed information, based on scenarios, about possible future global climate. Models focus on both the human side of the climate issue and on the natural system.

Models of the human system take into account how changes in economy, population, governance, and technology will change the drivers of climate change. Natural system models rely on numerical data and mathematical representations of physical, chemical, and biological processes.

Both types of models are equally important in documenting and explaining how our climate system will change over time. The value of these models has already been shown in the "Critical Experiment" discussed earlier in this chapter. Here we explain some very basic ideas about how those models are constructed and used.

Models of Human System: IPCC SRES Scenarios

The Special Report on Emissions Scenarios (SRES) identifies demographic change, social and economic development, and the rate and direction of technological change as the main driving forces of future GHG trajectories. We cannot foresee the future, and we cannot know how each of these driving forces will change in the future. A different approach is needed to characterize what is possible and what those possibilities imply about climate change. The approach now chosen is to identify storylines and scenarios that define the range of possibilities for each of these factors.

To represent a range of driving forces and resultant emissions, the SRES considers four "qualitative storylines," called "families," labeled A1, A2, B1, and B2 in Figure 2-8. From these four families, 40 alternative development scenarios are defined in six scenario groups (A1B, A1FI, A1T, A2, B1, and B2), which allow a qualitative representation of the impacts of policy choices that may favor a storyline.

Baseline GHG emissions scenarios are created from each of the 40 development scenarios. These emission scenarios drive models of the natural system that provide insight on implied climate changes and their likely impacts. The IPCC released six emissions scenarios in 1992, which provided modeling trajectories for GHG emissions on the basis of each development scenario through 2100. In 1996, these scenarios were revisited, and updated scenarios were presented in May 2000. There is also a baseline scenario that assumes no policy changes.

Changes in socioeconomic conditions are key in projecting changes in climate and emissions and also affect how or if mitigation or adaptation measures are feasible. As a hypothetical example, let's assume Region A in northern Africa has been noticing a steady decline in tourism. The primary draw to the region was observation of species migration patterns, but due to changes in climate, these patterns have been altered and are less predictable. Because of this, tourists are not willing to take the risk of traveling to see a migration that is not likely to occur, and the economic stability of the region declines. In doing so, the money available to invest in advanced technologies to accommodate tourists also decreases, people move to manufacturing jobs to recover lost income, and emissions increase, further contributing to global climate change. This is an example of the type of scenarios the SRES storylines seek to encompass.

Figure 2-8. SRES Qualitative Storylines for Modeling Scenarios

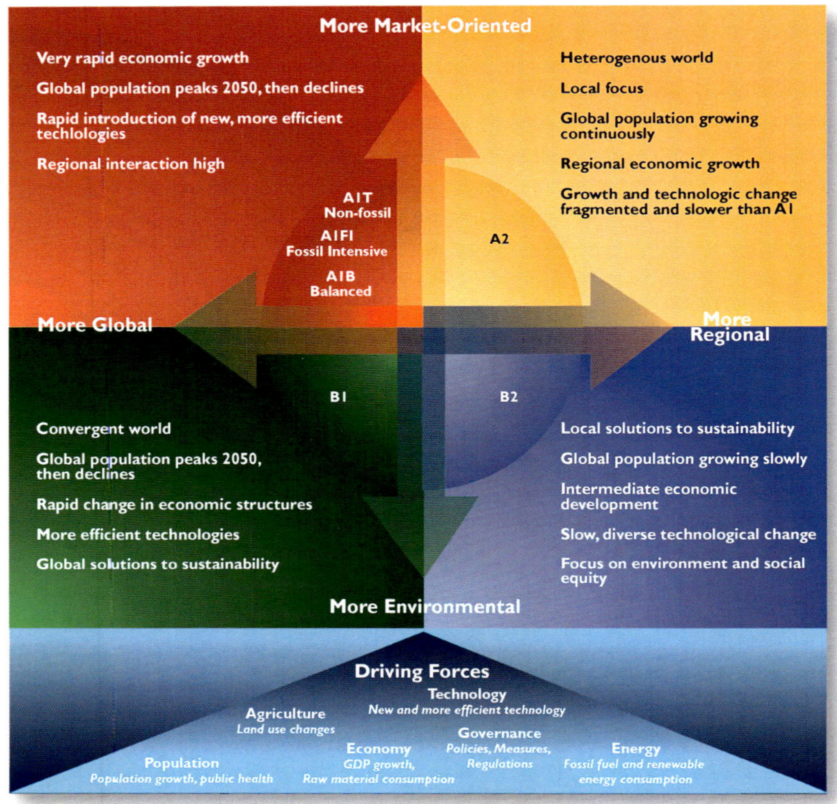

Note: GDP = gross domestic product.

Since SRES, other emission scenarios (called post-SRES scenarios) have been published, which are comparable in range but use lower drivers for emissions, such as population projections. The studies that have used the newer population projections have seen only slight changes in the projections of other drivers, such as economic growth.

Figure 2-9 shows global GHG emissions in 2000, and then compares projected emissions for 2030 using both SRES and post-SRES scenarios (using the terminology of Figure 2-8 to describe the scenarios). Emissions considered include CO_2, CH_4, N_2O, and the fluorine-containing GHGs.

The global surface warming (about 2°C) implied by these SRES scenarios is similar up to 2030 (Figure 2-10). After 2030, the modeled results diverge significantly. None show stabilization of temperature, but the temperature increase and rate of increase strongly depend upon the scenario chosen. The scenario representing stable emissions at the rate of the year 2000 shows the least and slowest temperature increase.

Figure 2-9. Global GHG Emissions for 2000 and Projected Emissions for 2030 using SRES and Post-SRES Scenarios

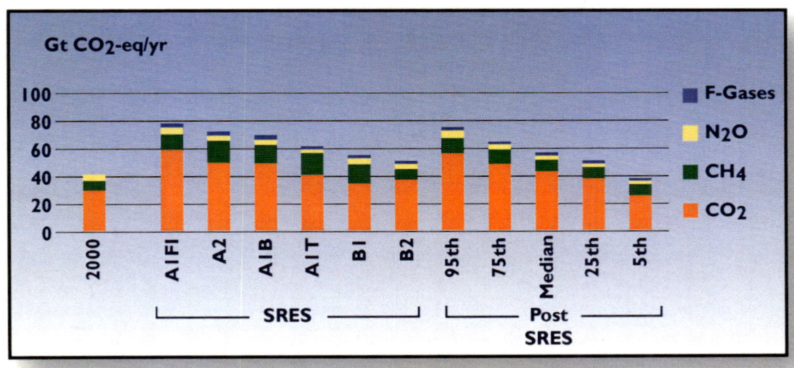

This set of experiments demonstrates the utility of the scenario approach in representing the effects of different policy choices and different development paths on climate. It also implies great difficulty in achieving a stabilization level.

> Progress since the IPCC Third Assessment Report enables an assessment that climate sensitivity is likely to be in the range of 2 to 4.5°C with a best estimate of about 3°C, and is very unlikely to be less than 1.5°C. Values substantially higher than 4.5°C cannot be excluded, but agreement of models with observations is not as good for those values.[1]

GHG concentrations must be stabilized at some acceptable level to reduce the effects of global warming. To do so, emissions must be stable (or decline), and concentrations would first peak and then decline over a period of many years due to the long atmospheric lifetimes of GHGs (see Table 1-1).

> Best-estimate projections from models indicate that decadal average warming over each inhabited continent by 2030 is insensitive to the choice among SRES scenarios and is very likely to be at least twice as large as the corresponding model-estimated natural variability during the 20th century.[1]

Figure 2-10. Global Averages of Surface Warming for SRES Scenarios

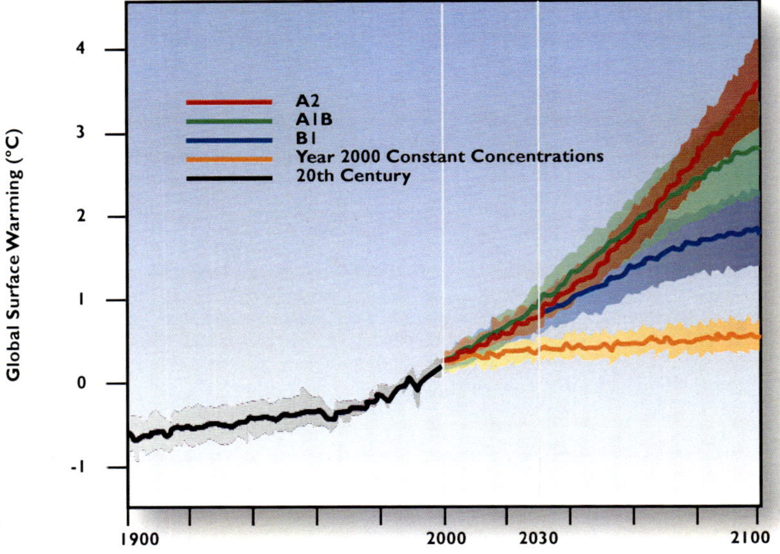

Note: The solid lines represent global averages of surface warming relative to 1980–99. Shading shows standard deviation ranges.

The lower the stabilization level, the sooner emissions must be stabilized. Mitigation efforts begun in the near future will have a large impact on opportunities to achieve lower stabilization levels. Chapter 4 discusses this further.

Models of Natural System

A quantitative climate model is a simplified mathematical representation of the earth's climate system. The ability of a model to simulate how the climate system will respond to a variety of inputs depends on how well the system is understood and the ability to integrate valid historical and observed data.

The more complex three-dimensional models are called general circulation models (GCMs) and are sometimes referred to as global climate models. These models include a three-dimensional representation of the atmosphere that simulates wind, moisture transport, and energy balance. These models reproduce clouds, wind, snow, rain, and most weather elements, allowing them to illustrate climate variability globally and regionally. Computations take significant time on high-end computer systems and provide coarse spatial resolution results.

> It is very unlikely that climate changes of at least the 7 centuries prior to 1950 were due to variability generated within the climate system alone. A significant fraction of the reconstructed Northern Hemisphere inter-decadal temperature variability over those centuries (7 centuries before the 20th century) is very likely attributable to volcanic eruptions and changes in solar irradiance.[1]

Figure 2-11 illustrates how GCMs have evolved since the 1970s as our understanding of the climate system has advanced along with technology. Through their nascent stage into the 1980s, they featured limited representation of the components of the climate system, which they modeled separately. As time went on, the processes and interactions of various parts of the system became better understood, enabling GCMs to incorporate and integrate more components.

Figure 2-11. Evolution of General Circulation Models

	1970s	1980s	Early 1990s	Late 1990s	Early 2000s	Mid 2000s
Atmosphere						
Land Surface						
Ocean & Sea Ice		Develop	Integrate			
Sulphate Aerosol			Develop	Integrate		
Non-Sulphate Aerosol				Develop	Integrate	
Carbon Cycle			Develop Land Develop Ocean	Combine Models	Integrate	
Dynamic Vegetation				Develop	Develop	Integrate
Atmospheric Chemistry			Develop	Develop	Develop	Integrate

Note: Components have been developed separately and later incorporated into the more sophisticated system model after they have been validated.

These models are able to model climate responses to various forcing agents, such as GHG, on a regional level. Models coupling the carbon cycle and other biogeochemical cycles are new and were not available for AR4.

All models have some error potential. Due to the sensitivity of the model as well as the variability and error potential among models, comparing multiple models and repeating analyses are necessary to represent the uncertainties in scenarios.

Coupled Dynamic System of People and Nature

As you can see, we need models of the human system to develop scenarios that give us guidance on how emissions will change over time. These models must incorporate economic, social, political, and technical knowledge to produce reasonable and believable scenarios.

The interactions in our globalized world mean that these models must represent whole sets of interlinked countries that have different structures of economic, political, and social behaviors. As one country adopts strategies, other countries are affected by trade patterns and other exchange. The systems are terribly complex and difficult to model in any comprehensive way.

You have also seen the awesome complexity of the natural system, where atmosphere, hydrosphere, cryosphere, oceans, soils, and biosphere interact at various time scales and with different lags and forces. The carbon cycle involves transfer of CO_2 (and CH_4) between various sources and sinks at different rates and efficiencies, and feedbacks in the system mean that changes in one sink or source affect other biogeochemical systems. Thus, prediction of how the carbon cycle will respond to a change in concentration of CO_2 in the atmosphere is very challenging. Our models are still evolving and only capable of coarse spatial resolution in computing climate under a given scenario.

However, trying to understand and predict the behavior of the human system and then using it to predict the response of the natural system are just the beginning. As the natural system changes, humans are affected, and human behaviors, economies, and political systems will change.

For example, we decide to use corn to make ethanol because we think it will reduce emissions. However, clearing forest land to grow corn removes an important carbon sink, so the reduction in emissions in our cities as we drive our cars on biofuel may not be as great as we think (and may not be a net reduction at all). Meanwhile, the albedo of the surface changes as the forest is removed, and the runoff from the fields is greater, which carries fertilizer into the ocean, and the chemistry of the ocean changes, and so on and so on.

The interactions, feedbacks, and cascade of influences are very real and ultimately determine whether the changes we make are positive. As we have learned, increased use of corn to produce biofuel may lead to higher prices for corn and stresses on food supplies in other, vulnerable countries. This may make it more difficult to mitigate climate change.

The network of connection is so tangled and complex, whether we can even imagine models capable of capturing enough to offer useful guidance is unclear. You will want to follow the development of such models and understand that they continue to improve. Much of the policy debate of the future will be guided by what these models suggest.

SUMMARY

The climate system consists of many dynamic components that produce natural variability in weather and climate regardless of external factors. Historical evidence shows that surface temperature has been increasing over the last few decades. Models show that observed changes over the past century can be reproduced if anthropogenic forcing is included, allowing scientists to conclude that an increase in GHG concentrations is the driving factor in rising surface temperature.

Recent increases in surface and ocean temperatures have caused changes in species and ecosystems, receding of glaciers, increased heat waves, changes in ocean salinity, and sea level rise.

We rely on models to illustrate how these systems will change in response to our decisions about emissions. We also rely on them to estimate how climate change may impact human and natural systems. They are a tool for informing policy discussions, and their results show how the entire global system is changing. Many countries contribute to the changes, most experience the consequences of the changes, and all will be affected by the solution.

This chapter introduces the basic science behind these actions. Many policies have been proposed to mitigate global warming, and Chapter 3 reviews the various national programs under consideration to reduce GHG emissions. Chapter 4 gives the available options for mitigation. Chapter 5 describes anticipated impacts of future climate change, and Chapter 6 presents steps that may allow your agency to adapt to those impacts.

---

[1] IPCC key finding. Throughout this book, each purple box shows a key finding from the IPCC with a high level of scientific understanding. See Appendix C for specific AR4 chapter sources for each finding. Chapter 1 gives a full discussion of our selection of key findings.

[2] Climate Change Science Program (CCSP) and the Subcommittee on Global Change Research, *The First State of the Carbon Cycle Report (SOCCR): The North American Carbon Budget and Implications for the Global Carbon Cycle*, Synthesis and Assessment Product 2.2 [King, A.W. et al. (eds.)], NOAA, National Climatic Data Center, Asheville, NC, www.climatescience.gov/Library/sap/sap2-2/final-report/default.htm.

[3] CCSP, 2008, *The effects of climate change on agriculture, land resources, water resources, and biodiversity in the United States.* A Report by the U.S. Climate Change Science Program and the Subcommittee on Global Change Research, by P. Backlund et al., U.S. Department of Agriculture, Washington, DC, USA, 362 pp., www.climatescience.gov/Library/sap/sap4-3/final-report/Synthesis_SAP_4.3.pdf.

[4] EPA, *The Science of Ozone Layer Depletion*, www.epa.gov/ozone/science/index.html, October 4, 2007.

[5] Simulations were completed using coupled Atmosphere-Ocean General Circulation Models. To achieve the results modeled in 2-4(a), 19 simulations were conducted using 5 models with only natural forcings, including volcanic activity. In 2.5(b), 58 simulations were run with 14 different models that included both natural and anthropogenic forcings. In both figures, the "model" line reflects the multimodel ensemble mean.

[6] EPA, *Future Sea Level Changes*, December 20, 2007, www.epa.gov/climatechange/science/futureslc.html.

[7] NOAA, National Climatic Data Center, *Global Warming Frequently Asked Questions*, www.ncdc.noaa.gov/oa/climate/globalwarming.html.

[8] IPCC, "Frequently Asked Question 3.3: Has there been a Change in Extreme Events like Heat Waves, Droughts, Floods, and Hurricanes?" in *Climate Change 2007: The Physical Science Basis. Working Group I Contribution to the Fourth Assessment Report of the Intergovernmental Panel on Climate Change*, FAQ 3.3, Cambridge University Press, Cambridge, United Kingdom, and New York, NY, USA, ipcc-wg1.ucar.edu/wg1/FAQ/wg1_faq-3.3.html.

[9] A. F. Hamlet et al., "20th Century Trends in Runoff, Evapotranspiration, and Soil Moisture in the Western U.S.," *Journal of Climate*, Vol. 20, No. 8, April 2007, pp. 1468–1486.

[10] Konrad Steffen and Russell Huff, *Greenland Melt Extent*, Cires Steffan Research Group, Boulder, CO, September 28, 2005, cires.colorado.edu/science/groups/steffen/greenland/melt2005.

[11] Anny Cazenave, "How Fast Are the Ice Sheets Melting?" *Science*, 314 (5803) November 24, 2006, pp. 1250–1252.

[12] To convert millimeters to inches, multiply by 0.03937.

Chapter 3
National Approaches to Managing GHGs

> There is high agreement and much evidence that a wide variety of national policies and instruments are available to governments to create the incentives for mitigation action. Their applicability depends on national circumstances and an understanding of their interactions, but experience from implementation in various countries and sectors shows there are advantages and disadvantages for any given instrument.[1]

The above IPCC finding suggests that a variety of policy instruments may be useful in avoiding adverse impacts of climate change. In this chapter, we describe the five major policy instruments for reducing GHG emissions currently proposed or available in the United States: regulatory initiatives, cap and trade, carbon taxes, voluntary measures, and technological fixes. Federal leaders with responsibility for climate policy should be familiar with all of them.

> *A variety of policy instruments may be useful in avoiding the adverse impacts of climate change.*

The United States is likely to use at least four, and possibly all five, of these policy instruments, and each is being applied in this country now, at least locally. Because they are currently in use, you will probably encounter them in your work and thus should understand them. We explain each policy instrument, show how to apply it, describe its current application, identify associated issues, and discuss how it can impact your agency.

As your agency is prompted to reduce emissions by increased costs or regulations, you will find the mitigation methods discussed in Chapter 4 useful. That discussion will more completely inform your assessment of costs of implementing these policies and your options in doing so.

The instruments discussed in this chapter are not mutually exclusive. For example, implementing cap and trade for CO_2 might require regulatory initiatives or guidance as industry seeks interpretations of various statutory provisions. Depending on the policy option selected, different agencies will be involved in developing, implementing, monitoring, and enforcing rules and congressional directives, which will affect the budgets of those agencies.

Our discussion focuses on the direct impacts of the policy on agency operation. Separately, in Chapter 5, we discuss the broader impacts of climate change if mitigation is not effective, some of which may bear on agencies. For example, federal public land managers will be concerned about the adverse impacts of global climate change on the ecosystems they manage, apart from the direct costs and benefits to their agencies, and significant costs can arise from the need for adaptation as described in Chapter 6. Additional international impacts are presented in Chapter 7, and these may require response by your agency.

These broader impacts are key considerations in a full cost-benefit analysis of these policies. A full assessment of the relative merits of the policies in terms of reducing potential adverse human and ecological impacts is beyond the scope or intention of this chapter but will be relevant to the choice among these five policy tools.

Regulatory Initiatives

The United States is considering a number of regulations to control GHG emissions.

Past environmental regulation has included specified pollution reduction technologies for desired emission reductions. Catalytic converters on motor vehicles are one example. Some governments are using regulations now to reduce GHG emissions, and the United States has imposed or is considering a number of regulations for this purpose. These include corporate average fuel economy (CAFE) standards, biofuel mandates, renewable portfolio standards, low-carbon fuel standards, building and appliance efficiency standards, and others.

CAFE Standards

The United States established CAFE standards via the Energy Policy Conservation Act of 1975. At that time, the purpose was to increase the fuel economy of the U.S. fleet in response to the Arab oil embargo of 1973–74.

At present, the fleet-wide average fuel economy of automobiles plus light trucks (which includes sport utility vehicles) is 21 miles per gallon (mpg). The Energy Independence and Security Act of 2007 (EISA) requires that new vehicles sold after 2020 achieve an average of 35 mpg, an increase of 40 percent. The National Highway Traffic Safety

Administration administers the CAFE program, and EPA measures the vehicle mileage.

For comparison purposes, Europe currently has a voluntary standard that vehicles average 40 mpg, which is implemented through taxes and other incentives. There are proposals to raise this to 49 mpg in the near future. Japan requires that vehicles sold there average 49 mpg.[2]

California sought to impose its own CAFE requirement through regulation of tailpipe CO_2 emissions. EPA did not accede to California's request, saying that the new congressional standard will achieve higher mpg than the state's proposal and that it did not have the authority to regulate GHG emissions. California claims that its standard will achieve 35 mpg in 2016, 4 years before the national standard. In November 2007, the state sued EPA in Washington, DC, District Court seeking to force it to render an official decision. The case went to the Supreme Court, and in April 2008, it found that EPA has the authority to regulate GHGs. Eighteen other states have committed to use the same standards as California. The EPA has released an advance notice of proposed rulemaking, and a decision is deferred.

Biofuel Mandates

Both the Bush administration (2000–2008) and Congress viewed biofuel mandates as a means to reduce petroleum consumption and GHG emissions while generating increased farm-sector economic activity and income.

The Renewable Fuel Standard is the primary regulation driving the production of biofuels in the United States. Established by the Energy Policy Act (EPAct) of 2005, amending the Clean Air Act (CAA), the Renewable Fuel Standard encourages the production of renewable fuels, including ethanol, biodiesel, and other motor vehicle fuels made from renewable sources. The Renewable Fuel Standard program as initially enacted established volumetric requirements for the production of renewable fuels in the United States, rising from 4 billion gallons in 2006 to 7.5 billion gallons by 2012.

In December 2007, EISA amended the Renewable Fuel Standard, increasing the renewable biofuel production requirements to 9 billion gallons in 2008, rising to 20.5 billion gallons in 2015, and 36 billion gallons in 2022. EISA also established a Renewable Fuel Standard limit of 15 billion gallons for corn-derived ethanol. Fuel sellers must derive the remainder of biofuel production from advanced and cellulosic biofuels (renewable fuels derived from cellulose, hemicelluloses and lignin, sugar, or waste products). Biofuels produced from these sources must show a reduction of 50 percent (advanced biofuels) or 60 percent (cellulosic biofuels) in GHG emissions relative to baseline fuels, such as gasoline, on a life-cycle basis.

The extent to which biofuels, such as ethanol from corn, reduce GHGs is controversial. The production and distribution of ethanol requires fertilizer, transport, and process energy. The use of fertilizer can result in higher N_2O emissions, a very potent GHG. Also, large increments in ethanol demand can bring land into production that previously was unused and absorbing GHGs. Although research at one time indicated about a one-third reduction in GHGs per gallon relative to baseline gasoline, more recent work suggests the reduction is much smaller and perhaps could even be negative.[3]

Renewable Portfolio Standards

Renewable portfolio standards apply to the power sector, where the policy requires a minimum percentage of renewable energy in a utility's portfolio of sources by a certain date. The power sector can meet this requirement with wind, solar, geothermal, small hydropower, landfill methane gas, or other sources.

Many states have adopted such standards. California requires its utilities to meet a 20 percent renewable portfolio standard by 2010, Connecticut has imposed a 27 percent standard for its utilities by 2020, Arizona 15 percent by 2025, and so on. In all, some 28 states and the District of Columbia have taken action toward setting a renewable portfolio standard, though in a few cases, such as Virginia, the state treats the standard as a voluntary goal, not a mandate.

A corollary of a renewable portfolio standard is to require users to purchase a specific percentage of their electricity from renewable sources. EPAct 2005 Section 203 requires federal agencies to purchase 3 percent from renewable energy sources in FY07–09, 5 percent in FY10–12, and 7.5 percent in FY13 and thereafter. This helps to create markets for renewable energy sources and thereby should reduce GHG emissions.

Attempts have been made to set a nationwide renewable portfolio standard in national energy legislation, but to date these have failed. The House version of EISA contained such a standard, but it was defeated in the Senate and not included in the final legislation. A cap-and-trade program is expected to greatly stimulate the development of renewable energy sources.

Low-Carbon Fuel Standard

A low-carbon fuel standard applies to motor fuels sold within a given jurisdiction. It mandates that these fuels meet a maximum carbon standard by a given date. The idea is to compel substitution of low-carbon fuels into the mix of fuels sold. These can take a number of forms, including biofuels, electricity (batteries), hydrogen, and natural gas.

To date, there is no national low-carbon fuel standard.[4] However, the state of California has established such a standard by EO and has issued a draft regulation. In that state, by 2020 suppliers must provide motor fuel with carbon content at least 10 percent less than petroleum-based fuels.

Building and Appliance Efficiency Standards

Appliance efficiency standards set minimum energy-efficiency requirements that household appliances must meet. The national program to set such standards has expanded greatly over the past 20 years. The National Appliance Energy Conservation Act of 1987 applied to dishwashers, refrigerators, washers, and dryers; fluorescent lamp ballasts were added later. EPAct 1992 added lamps, electric motors, and commercial heating and cooling equipment. Later, the act included room air conditioners, more ballasts, water heaters, residential central air conditioners and heat pumps, residential furnaces, and boilers.

Historically, the American Society of Heating, Refrigerating and Air-Conditioning Engineers (ASHRAE) has developed building standards in cooperation with DOE. ASHRAE is seeking to tighten its standards by 2010 to achieve a 30 percent improvement in energy efficiency relative to the current standard.

The U.S. Green Building Council Leadership in Energy and Environmental Design (LEED) is a third-party certification program for the design, construction, and operation of high-performance green buildings.

EPAct 2005 includes federal building sustainable performance standards. In 2006, OMB began using a new environmental scorecard for federal agencies, which includes a sustainable building element.

EO 13423 requires new construction and major renovation of federal agency buildings to comply with the *Guiding Principles for Federal Leadership in High Performance and Sustainable Buildings* set forth in the Federal Leadership in High Performance and Sustainable Buildings memorandum of understanding (2006). These principles are as follows:

- Employ integrated design principles.
- Optimize energy performance.
- Protect and conserve water.
- Enhance indoor environmental quality.
- Reduce environmental impact of materials.

EO 13423 specifically requires that 15 percent of the existing federal capital asset building inventory of each agency incorporate the sustainable practices in the guiding principles by 2015.

Most federal agencies have responded to these regulations with internal policies, and have implemented LEED certification for federal building design and construction.[5]

EISA Section 431 accelerates targets for energy use reductions in federal buildings to 3 percent per year for FY08–15, producing a 30 percent reduction in energy intensity by 2015, consistent with EO 13423.

EISA requires energy use reductions in federal buildings of 3 percent per year.

EISA Section 436 established an Office of Federal High-Performance Green Buildings in GSA for analysis, guidance, and training on life-cycle costing for green buildings.

Issues with Regulatory Initiatives

Some suggest that regulatory initiatives are necessary to correct market failures and that they work, reducing energy use and encouraging technological development. Further, such initiatives may be politically acceptable. There is a long record of their use at both the federal and state levels.

On the other hand, regulation may be more costly than other policy instruments and can have unintended consequences. For example, mandated efficiency standards force cheaper capital equipment (and buildings) from the market, raising costs to lower-income consumers. As a result, these consumers keep older, energy-inefficient equipment longer and have more trouble paying for new equipment when they do make a change.

Impacts on Agencies

A number of energy-related requirements have affected federal agencies already. Although not directly tied to GHG emissions, the requirements effectively reduce what otherwise would have been emitted.

EPAct 2005 requires federal agencies to reduce the energy intensity of their facilities, measured in British thermal units[6] per square foot of space, by 3 percent annually from a 2003 base year. Section 104 requires federal agencies to purchase ENERGY STAR and FEMP-designated energy-efficient products. EPAct 1992 requires at least 75 percent of the vehicles agencies acquire to be capable of burning alternative fuels beginning in 2000.

EO 13423 requires agencies to reduce the use of petroleum in their fleets by 2 percent per year and to increase the use of alternative fuels by 10 percent per year. It also encourages agencies to reduce GHG emissions but does not require them to do so.

EO 13432 requires DOT, DOE, and EPA to coordinate any regulations with respect to GHG emissions from vehicles and to consider science, benefits and costs, public safety, and economic growth. This EO immediately followed the Supreme Court decision (see "CAFE Standards") that gave EPA authority to regulate GHG emissions.

EISA requires agencies to reduce fleet petroleum use 20 percent by 2015 relative to 2003 and facility fossil fuel use 30 percent by 2015 relative to 2005.

Some in Congress are calling for federal agencies to lead by example and become low-carbon or carbon-neutral operations. EISA approved and funded a pilot program for carbon-neutral congressional operations.

To achieve these requirements, agencies will need to make significant investments in the replacement of vehicles, heating and cooling equipment, lighting, and other energy-using devices.

To achieve these requirements, agencies will need to make significant investments in the replacement of vehicles, heating and cooling equipment, lighting, and other energy-using devices. They also must formulate plans for ongoing investments and utilization of advanced technologies. In most cases, however, the investments should provide returns in the form of avoided energy costs. Thus, although these requirements stress budgets, the savings that accrue should help to ease the burdens.

CAP AND TRADE

Congress has studied and debated cap and trade and proposed a wide variety of legislation in the area. No national cap-and-trade program is now in place in the United States, although other countries do have them, as do regional groups, states, and cities in this country. We cannot be certain of the exact form of a future U.S. program, but we can draw a general picture on the basis of existing and proposed programs.

How Cap and Trade Works

A national cap-and-trade program would specify an annual limit (cap) on the quantity of GHGs that regulated entities could emit. The government would distribute allowances for emissions within a given period (such as a calendar year), either through public auction or specified formulas. To emit a ton of CO_2 legally, in the prescribed period, the emitter would be required to provide the government an equivalent allowance. The overall cap and number of allowances provided may be reduced over time.

Trading of allowances would be permitted and likely would take place on organized exchanges. The price of allowances would be determined by the demand for permits by CO_2 emission sources and supply of CO_2 emission permits issued by the government. Thus, a price for emission allowances would be established, which emission sources would take into account in determining their use of fossil and alternative fuels. Users of fossil fuels likely would face increased costs under a cap-and-trade program.

> There is high agreement and much evidence that: Policies that provide a real or implicit price of carbon could create incentives for producers and consumers to significantly invest in low-GHG products, technologies, and processes. Such policies could include economic instruments (cap and trade), government funding, and regulation.[1]

Sellers and buyers would make organized exchanges of allowances. Those able to curb emissions at a cost less than the market price could profit by selling allowances. For others, the cost of reducing emissions may exceed the cost of purchasing additional allowances. Cap-and-trade systems are intended to provide an economically efficient means to reduce overall carbon emissions by encouraging those who can cut emissions most cheaply to do so. Overall, cap and trade minimizes the costs of reducing GHG emissions while meeting programmatic objectives.

Since CO_2 is the dominant GHG, the other recognized GHG categories are translated into CO_2 equivalents on the basis of their global-warming potential (see Table 1-1).

A cap-and-trade program probably would specify the party responsible for monitoring GHG emissions and matching allowances to emissions. In existing pollution control programs, EPA requires emission sources to monitor and report their own emissions to EPA. This could be done for GHGs, although there are obvious trust issues. Someone must be responsible for ensuring that allowances tendered are bona fide.

Carbon offsets could be obtained either within the United States or internationally.[7] A party might obtain offsets domestically—from certain types of agricultural activities that absorb or sequester CO_2—or internationally through purchase of offsets on exchanges or through the United Nations (UN)–approved clean development mechanism and joint implementation projects.

A program might allow participants to bank allowances and save them for future use. In some cap-and-trade versions, "early action" credits could be earned for actions taken before the program officially starts. Participants also might be able to borrow from future years' allocations for use in the present. Cap-and-trade programs with this latter feature usually charge interest in the sense that a borrower would be required to pay back more than one allowance for each allowance borrowed.

Some proposed cap-and-trade systems also contain safety valve features, in which the government stands ready to provide additional allowances at a preset price. The purpose of this feature is to keep the price of allowances from rising too much over a short time. In addition, cap-and-trade programs can include a floor price, at which the government stands ready to purchase allowances to keep their price from becoming too low. One reason to maintain a floor price for allowances is to protect investments in GHG reduction projects.

Several other details surround the specifics of a cap-and-trade program, including which entities would have to submit allowances to cover their emissions, the scope of activities to be covered, and whether there should be exemptions for very small firms. Some proposals would chiefly include utilities, oil refiners, natural gas processors, and oil and gas importers, with possible exemptions for very small entities in these industries. Others would include large industrial users of fossil fuels, while some would leave the design of the system to the administrator.

For illustration purposes, we present GHG emission profiles under a variety of recent cap-and-trade proposals from the 110th Congress (Figure 3-1). As the figure illustrates, all of the proposals would reduce U.S. emissions below business-as-usual projections. However, some would do so much more rapidly and some (such as Sanders-Boxer) set a goal related to *GHG concentration stabilization* although the stabilization level is at least 170 ppm above natural values (see Figure 1-3) and presumes compatible actions of other countries. Others may achieve *emissions stabilization* under which concentrations would continue to increase.

Figure 3-1. Comparative Effects on Emissions (to 2050) of Proposals in 110th Congress

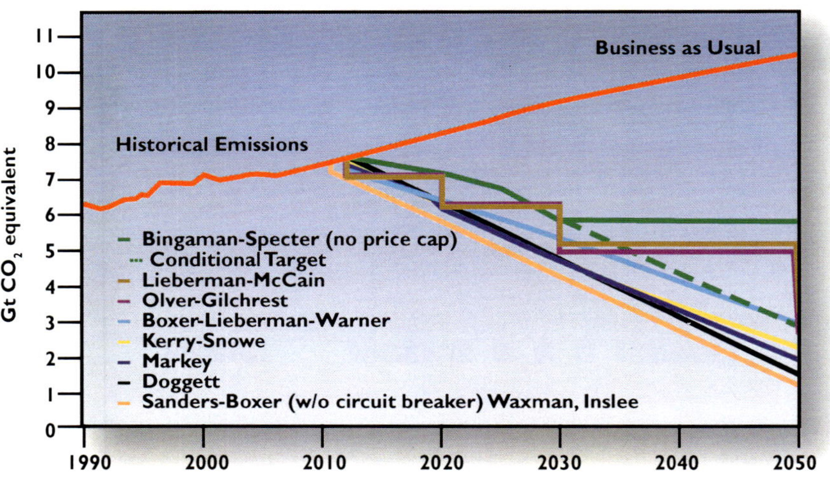

Table 3-1 compares per capita covered emissions under a business-as-usual forecast to those under one of the leading legislative proposals in the 110th Congress.[8]

Table 3-1. U.S. Per Capita Covered GHGs

Year	Covered emissions[a] (Mt CO_2-eq)	U.S. population (million)	Per capita GHG emissions under business-as-usual	Per capita GHG emissions under S.2191
2006	5,661	299.4	18.9	18.9
2012	5,995	314.3	19.1	16.6
2020	6,614	335.8	19.7	13.2
2030	7,783	363.6	21.4	9.6

Note: Mt CO_2-eq = million tons of CO_2 equivalent.

[a] Covered emissions are assumed to be 80 percent of total U.S. GHG emissions.

As seen in Table 3-1, per capita emissions would drop 30 percent by 2020 and drop 50 percent by 2030. Unless non-fossil-energy sources largely replace fossil over the period, people would have to use considerably less energy than they do now.

Existing U.S. Programs

Under the CAA Amendments of 1990, a cap-and-trade program involving power plant sulfur dioxide (SO_2) emissions began in 1995. Initially, 110 plants were involved, but by 2000, virtually all of the nation's power plants were included. The CAA Amendments formed an organized exchange in which organizations traded SO_2 emission allowances. By one estimate, the SO_2 trading program saved $3.5 billion annually over the cost of a regulatory approach to achieve the same outcome.[9]

A nitrogen oxide (NO_x) emissions trading program began in 1999 under the Ozone Transport Commission, a multistate organization created under the CAA. EPA now operates the NO_x Budget Trading Program, covering 21 states plus the District of Columbia, between May 1 and September 30 of each year. States divide targets among emitters, who can trade NO_x emission allowances with one another to achieve the overall goals. Over time, the number of NO_x allowances has declined, exerting upward pressure on allowance prices and decreasing the contribution of NO_x to smog in the eastern part of the United States.

In 1994, the South Coast Air Quality Management District in Southern California initiated a tradable emission permits program called RECLAIM (Regional Clean Air Incentives Market) to curb the costs of reducing smog in the Los Angeles basin area. The initial program covered emissions of N_2 and sulfur oxides (SO_x) from stationary sources. More than 350 sources participate in the NO_x market program and 40 in the SO_x market. A few firms have volunteered to participate in RECLAIM, presumably because they can curb their emissions more cheaply than by buying permits and can profit by selling their unneeded emission allowances to other emission sources.

The CCX system is voluntary but legally binding.

The Chicago Climate Exchange (CCX) now operates a voluntary but legally binding system in which members agree to reduce carbon emissions below an agreed-upon baseline. Reductions below that baseline can be sold in the CCX or banked. Those who emit above the baseline must purchase these allowances. CCX began operations in 2003 and has about 300 members, including at least 7 U.S. municipalities and 2 states.

European Trading System

To meet its targets under the Kyoto Protocol, the European Union (EU) created a cap-and-trade program in 2003 called the European Trading System; the first 3-year phase ran from 2005 through 2007 and the second is now underway (2008–2012). This program includes some 12,000 firms in six major industrial sectors that are responsible for monitoring and reporting their own emissions and for submitting sufficient allowances to cover their CO_2 emissions.

Under the European Trading System, each emitter receives allowances for emissions. In the first phase, they were provided to firms at no cost, although there is movement toward auctioning in the future. In the second phase, allowances are reduced by 7 percent below 2005 emission levels. Trading has taken place on organized exchanges, such as the European Climate Exchange. The European Trading System has dealt only with CO_2 emissions to date, but inclusion of N_2O and PFCs has been proposed but not yet finalized. During the first phase, emissions by regulated entities rose by 1.9 percent.

In the first phase, the EU treated the European Trading System as an experiment, with emission caps set separately by each country but subject to European Commission approval. However, the result was an overallocation of allowances relative to need. When market participants discovered this, the price of allowances plunged, creating instability and widespread uncertainty. However, in its current phase, the reduced national European Trading System allocations have increased the market price for allowances.

Figure 3-2 illustrates the upward trend over time in the market price for allowances. The figure shows what investors expected the price of an allowance to be in December 2008 (the December 2008 futures price) over the period January through early July 2008. The price of such an allowance was as low as 19 euros at the beginning of February but had reached as high as 29 euros in early July. This illustrates how the market responds as tighter rules are considered. In general, cap-and-trade systems are designed to reduce allowances over time, raise the cost of emissions, reduce use of fossil fuels, and lower emissions.

Figure 3-2. December 2008 Futures Prices for One Allowance Unit (1 Ton CO_2) in January–July 2008

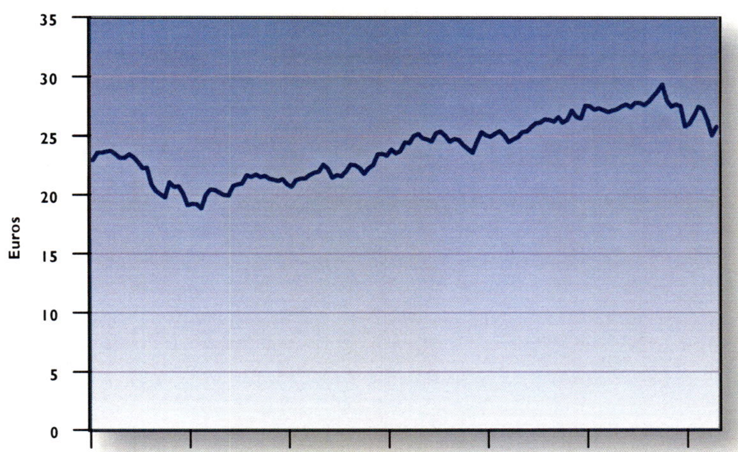

Regional Programs

Several regional cap-and-trade programs for CO_2 and GHGs are underway. These involve regions of the United States alone or in combination with Canadian provinces and Mexican states.

The Regional Greenhouse Gas Initiative (RGGI) covers CO_2 emissions from power plants in 10 Northeastern and Mid-Atlantic states.[10] RGGI sells all allowances each quarter at auctions that began in September 2008.

The Western Climate Initiative (WCI) involves seven western states and four Canadian provinces: Manitoba, British Columbia, Quebec, and Ontario. Other U.S. states, Canadian provinces, and Mexican states participate as observers. The WCI covers multiple sectors and involves a registry of emissions that crosses national borders. The goal is for participants to reduce their GHG emissions by 15 percent relative to 2005 by 2020.

Members of the Midwestern Greenhouse Gas Accord, signed in late 2007, include six U.S. states and the Canadian province of Manitoba plus additional observers. It sets up a cap-and-trade program involving multiple sectors to enable states to reach individually set targets. The accord also establishes a system for tracking the GHG emissions of entities responsible for reductions.

If a national cap-and-trade program is established, the existing regional, state, and municipal efforts may eventually combine with the national program. Programs that include emissions from users of fuel are more complicated than those that focus on sellers. Firms may prefer a single program rather than multiple targets.

With a single national program, regions can avoid the costs of administration. In the meantime, regional cap-and-trade markets will provide insights into some of the challenges in making such a market work smoothly at the national level.

Issues

Many policymakers support a cap-and-trade program for GHGs because it offers a cost-effective way to achieve targeted reductions. Allowances can be distributed or auctioned, and the latter provides revenues that can be used to administer the program and develop low- or no-carbon energy alternatives or for other purposes. A cap-and-trade program also appears politically feasible on the basis of the success of this approach in reducing SO_x and NO_x emissions.

However, operating a cap-and-trade program poses several challenges. It requires monitoring by government and private emitters. Because it may be international in scope, such monitoring could involve ensuring that the emission offsets obtained from abroad are legitimate.[11]

Monitoring costs can be reduced by requiring emission sources to report their emissions to the government, although the government would need to periodically check the accuracy of the reports, which is done for many pollutants regulated by EPA. For example, power plants monitor other emissions under the CAA.

An important issue with cap and trade is how fast to reduce the number of allowances to reach the cap.

An important issue with cap and trade is how fast to reduce the number of allowances to reach the cap. A system that decreases allowances relatively rapidly would enable the country to achieve reduction goals more quickly, but might exert greater upward pressure on fossil fuel prices and the cost of emission reduction, with consequent effects on energy costs and the economy. If the U.S. population continues to rise, rapid decreases in allowances would force substantial reductions in per capita use of fossil energy, which in turn would necessitate substantial changes in personal lifestyles within a short period.

Although many legislative proposals for emissions trading permit the use of offset credits, whether a U.S. cap-and-trade program would allow their use is uncertain. Furthermore, the types of offsets permitted under cap-and-trade proposals vary: some allow for international offsets, but others do not.

Linking cap-and-trade programs within this country or with those abroad presents challenges that differ slightly from ensuring the integrity of offset credits. For example, would a carbon allowance under a U.S. program be measured in the same way as an allowance under the European Trading System?

Energy price volatility and rapid energy price increases from a cap-and-trade program could harm the U.S. economy. However, the higher fossil fuel prices under cap and trade should reduce the demand for these fuels as emission sources switch to alternative fuels that generate few or no GHGs (such as nuclear, solar, wind, or ocean energy). Higher prices could decrease oil imports and exert downward pressure on world oil prices, ameliorating the upward pressure on these prices from the cap.

Finally, the distribution of free allowances and the spending of monies gained through auctioning allowances will spur private and public entities to secure them, potentially raising questions of fairness and political favoritism. However, if all allowances were auctioned, the problem of distributing free allowances would disappear.

Impacts on Agencies

The most direct impact on agencies of cap and trade would be an increase in the cost of energy derived from fossil fuels. This could include motor fuels and purchased electricity, the amount depending on the specifics of the cap-and-trade program and possibly increasing as the cap tightens over time.

Energy cost increases will directly affect agency budgets; budget planning will need to take these cost increases into account.

Budget planning may need to consider energy price volatility (such as might arise through volatility in emission allowance prices as shown in Figure 3-2). Agencies may need contingency funds to deal with higher-than-expected energy costs. Including the means to deal with price volatility in a cap-and-trade system would reduce the degree of volatility and therefore its effects.

Higher energy costs would affect agency choices for vehicle replacement and facility recapitalization. Higher vehicle fuel economy will become more attractive compared with other considerations. Similarly, investment in facility efficiency that previously appeared uneconomic will become more attractive as energy costs rise.

Also, higher energy costs will likely affect agency operating procedures. Agencies that deal with energy more directly would be most affected. DOT, which relies on gas tax receipts and user fees for funding of highway construction and transit, could expect revenue to fall under cap-and-trade or carbon taxes, other things equal.

DOE and EPA could also be affected by changes in the quantity and type of fuels consumed. For example, as fuel costs rise under cap and trade, an agency's number of vehicles and miles driven likely will come under greater scrutiny, and the result may be fewer authorized trips. If prices rise, transit ridership will increase in metropolitan areas, which will increase congestion for mass transit as well as increase maintenance cost for a more-burdened transit system.

Federal leaders will be interested in the estimated costs of the GHG emission reductions detailed in Chapter 4.

A cap-and-trade program may also offer agencies certain opportunities. Some agencies may be able to produce carbon offsets for sale or trade. For example, agencies that manage public lands may be able to use those lands to add carbon-absorbing vegetation that qualifies for emission credits.

Some agencies may be able to take advantage of the revenues raised by auctioning off allowances. Programs to promote energy efficiency and low- or no-carbon forms of energy appear to be good candidates to receive increased budget allocations from such monies. Thus, agencies that sponsor these programs—either to manage their own GHG emissions or on behalf of the constituencies they serve—may be able to take advantage of new funding opportunities.

A national cap-and-trade program might span all major aspects of the economy, probably making it harder to design and implement. In addition, Congress may not be able to write a law prescriptive enough to address all the domestic and international issues arising under cap and trade.

Instead, Congress may charge agencies with crafting rules to address these details (monitoring, offsets, and regulations for carbon capture and sequestration, for example). As a result, agencies with the authority to run and enforce these programs may have to flesh out detailed regulations and guidance documents through agency rulemaking conducted under the Administrative Procedure Act. In addition, new control standards may be necessary over time to stay current with changing technologies and a changing climate as well.

An agency required to draft numerous rules and enforce congressional mandates will require significant resources to hire staff members. Given that many rules under the CAA have a history of litigation, more attorneys, scientists, and economists might be needed to address implementing these rules and directives in a timely fashion. All of these will require increasing agency budgets.

CARBON TAX

Another way to control GHG emissions is to impose a tax on the carbon content of fossil fuels. Some economists have asserted that a carbon tax is an efficient way to force emitters to bear the costs they impose on others.[12] They argue that a carbon tax would cause less energy price volatility than a cap-and-trade system, it would be simpler to administer, and the revenues could be used to reduce other taxes, such as those on capital or labor. In their view, we reduce taxes on things we want and increase them on things we don't.

How would such a tax work? A carbon tax would be imposed on coal, oil, and natural gas according to the carbon content of each of these fuels.[13] It would fall most heavily on coal, next on oil, and least on natural gas. The basic notion would be to impose it on producers of fossil fuels or first purchasers, in the case of imports. They in turn would pass a portion of the tax on to utilities in the case of coal, utilities or industrial customers in the case of natural gas, and refiners or distributors of imported products in the case of oil. Consumers would experience the tax in the form of higher prices for electricity, gas, and oil and of goods and services that utilize these energy inputs in their production processes. Higher prices for these energy sources would reduce their use, other things equal. This is similar to the way an upstream cap-and-trade program would work.

Chapter 4 discusses the relation between the price of carbon emissions and the economic potential for emissions reduction by technological improvements. Such an analysis provides one means to set the price of carbon in a carbon tax.

Users of fossil fuels could reduce their consumption to avoid the tax or, if such reduction is more costly than the tax, pay it and treat the tax as a cost of use. Over time, if the potential costs of climate change increase, the tax on GHG emissions presumably would rise accordingly.

Refiners of oil products may pass a portion of the tax on to motorists, airlines, and other customers. In this way, the cost of gasoline, diesel, jet fuel, and other products would increase. Motorists would be given incentive to drive less and to purchase more fuel-efficient vehicles. The carbon tax could stimulate the use of alternative fuels, such as ethanol derived from renewable sources, and of hybrid electric vehicles that use both batteries and fuel for propulsion.

Energy consumers would pay a portion of a carbon tax. In the process, the relative prices of fuels would change: coal would become more expensive relative to oil and natural gas, and oil more expensive relative to gas. Utilities might pass the tax through to their customers, but could avoid it to the extent they substituted nuclear and renewable forms of energy for coal, oil, or gas.

Current and Proposed Taxes

Finland enacted a carbon tax in 1990, but subsequently transformed it into a combination carbon and energy tax. At present, the tax rate is 20 euros per ton of CO_2, or approximately \$25 per ton of CO_2 in dollars, or \$95 per ton of carbon.[14]

Sweden enacted a carbon tax in 1991 and currently taxes CO_2 at a rate of \$150 per ton, but it does not apply the tax to fuels used to produce electricity and exempts industry from half the tax.

Sweden currently taxes CO_2 at a rate of \$150 per ton.

The United Kingdom imposed a "climate change levy" in 2001 on the use of nonrenewable energy in the industrial sector (households and transport are exempt). Tax rates vary by fuel; the rate on electricity is £0.00456[15] (\$0.0066) per kilowatt-hour (kWh) (on nonrenewable fuels, including nuclear), which would be about 10 percent of the average cost of electricity in the United States. The program is accompanied by a reduction-trading scheme, so those who reduce below goals reduce their tax burden and can sell those reductions for others to use as credits. Britain redistributes tax revenues from the climate tax to private-sector employers in the form of reductions in their National Insurance contributions. This imposes no net tax increase on British industry.

The Canadian province of Quebec added a carbon tax in fall 2007. Because hydroelectricity accounts for most of the province's power, the tax essentially falls on petroleum products. The rate on these products is initially quite low, only 4¢ per gallon in U.S. dollars.

British Columbia's revenue-neutral $10-per-ton carbon tax took effect on July 1, 2008, and applies to all fossil fuels. The rate increases by $5 each year until it reaches $30 per ton. The tax is broadly applied, but there is protection for lower-income families.

In April 2007, the city of Boulder, CO, enacted a tax on electrical energy sold by the local utility within city boundaries. This is sometimes called a "carbon tax" because most of the electricity used in Boulder comes from burning coal. The tax rate is equivalent to about $7 per ton of carbon.

The San Francisco Bay Area Air Quality Management District (covering nine counties in the Bay Area) carbon tax of 4.4¢ per ton of CO_2 took effect in July 2008. Businesses are required to measure and report their own use of carbon. Proceeds are used to fund the management district's emissions-monitoring program.

The 110th Congress featured two proposals to impose carbon taxes.[16] Congressman John Dingell of Michigan discussed a carbon tax to phase in over 5 years to $50 per ton of carbon and then be adjusted for inflation. In addition, his proposal included a 50¢ per gallon tax on gasoline and jet fuel, also phased in over 5 years and then adjusted for inflation.

In most of these instances, the tax rate is so low that it likely will have little effect on behavior. For example, in the Bay Area program mentioned above, the typical business pays $1. Further, despite these isolated examples, carbon taxes are not in widespread use.

Issues

A carbon tax is a less certain way to reduce GHGs than cap and trade.

A carbon tax is a less certain way to reduce GHGs than cap and trade. If the demand for fossil fuels is inelastic (not very sensitive to price), it may take a very high tax to spur significant GHG reductions. Such a high tax may have unforeseen and unintended consequences.

An optimal carbon tax would accurately reflect the costs imposed by GHG emissions. Because many of the costs of current emissions come in the form of future impacts (Chapter 5) and adaptation strategies (Chapter 6), establishing such costs with confidence is challenging. The *Stern Review* is an example of such an effort, though the approach it takes has been questioned.[17]

As carbon control technologies and hence the cost of reducing carbon emissions changes, the tax would have to be adjusted. Also, a carbon tax would have to escalate over time to generate ever-increasing reductions in carbon emissions. This could create political difficulties and eventually harm the economy.

Harmonizing carbon taxes among countries is a challenge. If such taxes are higher in some jurisdictions than in others, then the low-tax countries could gain an artificial comparative advantage in the production of energy-intensive goods.[18]

Finally, some question the political feasibility of any form of carbon tax. The American public generally opposes tax increases and might not accept the magnitude of carbon tax necessary to reduce U.S. GHG emissions to the extent implied by the costs of such emissions.

Impacts on Agencies

A carbon tax could act in much the same way as cap and trade, raising the cost of petroleum products and electricity, thus increasing energy costs to agencies and impacting their budgets. The volatility of energy prices probably would be less under a tax, however, reducing that source of uncertainty for agency budgets. Long-range budget planning would have to factor in any periodic adjustments in the tax to reflect the marginal cost and marginal benefit of reducing GHG emissions.

The increased cost of energy would affect agency investment decisions concerning vehicles and facilities. As with cap and trade, the higher energy costs would yield higher returns on investment for fuel-efficient vehicles and energy-efficient facilities. It also would drive agencies to examine means of reducing energy use through changes in operating procedures, such as fewer employee trips or the like.

Even a modest carbon tax would raise very substantial revenue.[19] If the U.S. government used a share of the revenue to support federal energy-efficiency and low-carbon energy technology programs, agencies sponsoring such programs could obtain increased support.

Voluntary Programs

Over the past 20 years, the United States has largely relied on voluntary programs to achieve carbon reduction goals. Such programs take several forms: reporting programs, public-private partnerships,

> *Over the past 20 years, the United States has largely relied on voluntary programs to achieve carbon reduction goals.*

and government provision of new energy-efficiency technology and information to the private sector. Although the United States has been criticized for not taking sufficient steps to reduce GHG emissions, over time its voluntary programs have helped to reduce the GHG intensity of its output—GHGs divided by GDP—faster than most other major countries in the Organization for Economic Cooperation and Development (OECD).[20]

Reporting Programs

Under EPAct 2005 Section 1605(b), DOE has maintained a "climate registry," under which firms can voluntarily report their GHG emissions and the steps they are taking to reduce them. According to DOE, more than 200 utilities, industrial firms, and other institutions now report their GHG emissions annually.

Under the Omnibus Appropriations Act of 2008, EPA is required to establish guidelines for companies to report their GHG emissions. EPA was required to develop proposed guidelines by fall 2008 and final guidelines by June 2009. For those covered under EPA's reporting rules, the reporting will be mandatory, but for the present, they are not required to do anything more than supply this information publicly.

A multistate climate registry for GHG emission reporting has formed.

At the state level, a multistate climate registry for GHG emission reporting has formed, which includes 39 U.S. states as well as 7 Canadian provinces and 6 Mexican states. In April 2008, the registry reached agreement on a set of standards governing the measurement and reporting of GHG emissions.[21]

Public-Private Partnerships

Several federal agencies have developed programs under which private-sector entities voluntarily pledge to reduce their GHG emissions. The firms negotiate the extent and timetable for reductions with the agencies and receive public commendation if they meet targets.

EPA sponsors the Natural Gas STAR, Coalbed Methane Outreach, and WasteWise programs and the trifluoromethane partnership. Each of these programs involves partnerships between EPA and private-sector firms that voluntarily agree to reduce GHGs via application of advanced technology or changes in processes. For example, in Natural Gas STAR, EPA partners with the oil-and-gas industry to reduce CH_4 emissions during production, processing, transport, and distribution operations, while in Coalbed Methane Outreach, EPA partners with coal mining firms to capture CH_4 and use it profitably rather than allow it to escape into the atmosphere.

The EPA Green Power Partnership is a voluntary program that supports the purchase of green power (solar, wind, geothermal, low-impact biomass, and low-impact hydro) with expert advice, technical support, tools, and resources. Using green power through this partnership allows an agency to lower its transaction costs, reduce its carbon emissions, and communicate its leadership to key stakeholders. One of the largest users of renewable power among the partners is the U.S. Air Force, which bought 9 percent of its electrical power through this partnership in FY08. EPA purchases 100 percent of its power in this way.

One of the largest users of renewable power is the U.S. Air Force.

Climate Leaders is an EPA industry-government partnership that works with companies to develop comprehensive climate change strategies. Partner companies commit to reducing their impact on the global environment by completing a corporate-wide inventory of their GHG emissions based on a quality management system, setting aggressive reduction goals, and annually reporting their progress to EPA. Through program participation, companies create a credible record of their accomplishments and receive EPA recognition as corporate environmental leaders.[22]

In 2004, EPA launched SmartWay—a brand that represents environmentally cleaner, more fuel-efficient transportation options.[23]

DOE sponsors similar types of programs, among them Climate Challenge, Regional Carbon Sequestration, Clean Cities, and Best Practices. Climate Challenge and Regional Carbon Sequestration are aimed directly at ways to reduce GHGs. Clean Cities is targeted at encouraging alternative motor fuels and fuel economy, while Best Practices is targeted at ways to conserve energy use. An indirect effect of these two programs is to reduce GHGs as well.

Some voluntary GHG reduction programs are managed jointly by one or more agencies. For example, DOE and EPA jointly manage Climate Wise, a private-public partnership in which firms commit to reduce their GHGs in exchange for public recognition, and Combined Heat and Power, a program aimed at more efficient forms of energy production. DOE, EPA, and USDA jointly manage AgStar, a program to capture CH_4 emissions in agricultural activity and to retain carbon in the soil. DOE,

EPA, DOT, and USDA jointly sponsor ClimateVision, a voluntary public-private partnership initiative to improve energy efficiency and GHG intensity in energy-intensive industrial sectors. Although there is disagreement over how much reduction such voluntary programs achieve,[24] the administering agencies indicate that emissions are significantly lower than otherwise. The 1997 *Climate Action Report* submitted by the United States under the UN Framework Convention estimated that the voluntary programs of DOE and EPA alone would result in a reduction of about 9 percent in net U.S. GHGs by 2010.

Information Provision

Simply providing information to the private sector, including consumers, is another way to accomplish GHG reduction. EPA's ENERGY STAR program probably is the best-known example. EPA sets energy-efficiency standards that manufactured products, such as appliances and home electronics, must meet, and places an ENERGY STAR label on those that qualify. ENERGY STAR extends to new buildings, including residences, and offers guidelines to consumers concerning how to increase the energy efficiency of existing homes. Within the program, more than 1,000 manufacturers offer products covering more than 30 commercial and residential product categories, including light bulbs, televisions, electronics, and telephones.

Voluntary National Goal

In 2002, President Bush announced that the United States would seek to reduce the carbon intensity of its GDP 18 percent by 2012. In 2002–06, the United States experienced rapid economic growth, and the carbon intensity of its output dropped by almost 10 percent. The contribution of officially sponsored voluntary programs is an important part of the efforts Americans have made to reduce their emissions relative to GDP.

Issues

Voluntary programs alone cannot reduce U.S. GHGs enough to avoid significant climate change.

Voluntary programs to reduce GHGs have had success at low cost. Such programs enable a wide choice of technologies and operational procedures, while still allowing firms to increase the goods and services they offer to consumers. Thus, governments and organizations participating in voluntary programs can make efforts to reduce GHGs while growing GDP. Participating governments can then use a share of this GDP growth to finance various social goals, including climate-related objectives.

Nevertheless, voluntary programs alone cannot reduce U.S. GHGs enough to avoid significant climate change. The IPCC has concluded that the full suite of existing programs is not enough to reduce GHG emissions.

> There is high agreement and much evidence that with current climate change mitigation policies and related sustainable development practices, global GHG emissions will continue to grow over the next few decades.[1]

A recent (2008) projection by DOE underscores this finding. Figure 3-3 displays world GHG emissions projections under business-as-usual assumptions through 2030 for countries in the OECD, and those outside. In 2005–30, projections show growth of about 28 billion tons to about 42 billion, or by 50 percent.

Figure 3-3. DOE Projection of Worldwide Carbon Emissions under Business-as-Usual Assumptions

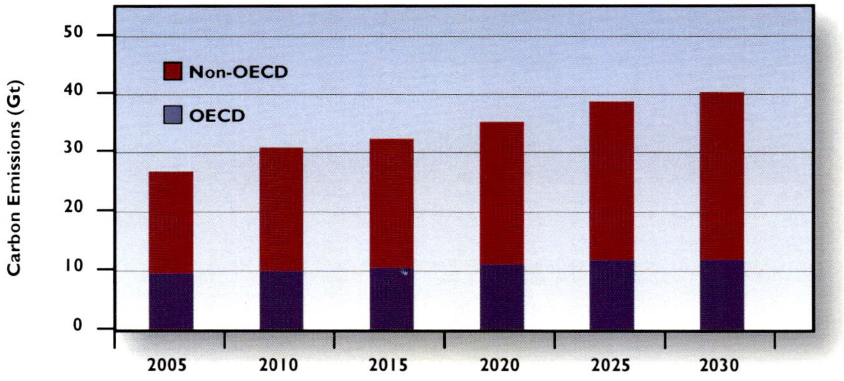

Impacts on Agencies

Voluntary programs offer agencies a variety of opportunities. An agency can itself pledge to meet a target level of emissions, perhaps in partnership with an environmental organization, and gain favorable publicity by meeting its goal. Because agencies already are compelled indirectly to reduce their GHGs through increased vehicle and facility efficiency, meeting such a target may not be much of an additional burden. Of course, a "voluntary pledge" at headquarters may be a requirement at offices in the field. The operational effect on managers may be indistinguishable.

An agency may be able to form a public-private partnership with some of its constituencies, cooperatively setting GHG targets with the constituents and providing favorable publicity when the targets are met. Such partnerships require funding support, so agencies will have to factor them into budget planning.

An agency can disseminate practical information to its constituencies about how to reduce GHGs (such as through its website or other means). Every agency's constituencies produce GHGs, so solidly based and clearly presented information should prove useful.

Agencies have other options as well. They can encourage private-sector groups with which they deal (trade associations, for example) to establish voluntary GHG reduction programs so that all of their members are included rather than just those already doing so as individual companies. They can participate in those groups' standard-setting processes to try to move the standards toward more energy-efficient levels. Moreover, they can encourage companies in their constituencies to label their products with the amounts of GHGs emitted in producing them. No single one of these efforts will have a huge effect on the nation's GHG emissions, but collectively they can make an important contribution.

Technological Fixes

> There is high agreement and much evidence that: The economic potential in the industrial sector is predominantly located in energy-intensive industries. Full use of available mitigation options is not being made in either industrialized or developing nations.[1]

Some believe that present technology is insufficient to achieve large-scale reductions in carbon without severe economic dislocation. They suggest that rapid efforts to decrease U.S. GHGs will lead to substantial economic cost, reducing people's well-being, and ultimately reducing political support for GHG reduction programs.

To avoid this, they propose that the United States invest substantial sums in new energy production and energy-efficiency technologies to reduce the cost of managing carbon and achieving carbon-reduction goals. This course of action raises questions concerning how much to spend on these technologies, how to fund such an investment program, and which technologies to support. It also raises the question of how public investment of this type will interact with private investments.

> There is high agreement and much evidence that: Government support through financial contributions, tax credits, standard setting, and market creation is important for effective technology development, innovation, and deployment. Transfer of technology to developing countries depends on enabling conditions and financing.[1]

Existing Programs

The U.S. government spends about $3 billion annually on energy R&D. This funding, 85 percent of which DOE manages, is focused on improved production technologies and energy efficiency. This funding supports new technology and encourages its use by the private sector. Recent technological breakthroughs have occurred in thin film solar technologies, wind turbines, biofuel production, superconductivity, lithium batteries, industrial process catalysts, and steam turbines. Each alone has a limited effect on U.S. GHGs, but taken together they achieve significant reductions while increasing productivity in the economy.

Opportunities

The U.S. Climate Change Technology Program (see Chapter 1) supports carbon-reducing energy production ranging from substantial new investment in nuclear power to capturing solar energy in space and transmitting that energy for use on earth. Other candidates include carbon capture and storage at power plants, removing carbon from the air, fuel cells to produce power and provide mobility, direct hydrogen usage, biofuels (including ethanol from cellulosic material and fuels derived from algae), ocean energy, vastly more efficient solar photovoltaics, and economy-wide use of lightweight hybrid vehicles that utilize alternative fuels, such as methanol or ethanol, and run in part on batteries.

Multiple GHG technologies must reduce emissions in greater and greater amounts over time to achieve the stabilization levels described in Chapter 2. Use of known technologies, such as biofuels, energy efficiency, and nuclear power, can reduce the world's GHG emissions rate (Figure 3-4). The wedges in Figure 3-4 represent the potential to reduce the world's CO_2 emissions in increasing amounts over time from eight available technologies. With application of all eight technologies, each increasing its contribution over time, world carbon emissions could remain flat for the next 50 years.[25]

Figure 3-4. GHG Emission Stabilization Wedges

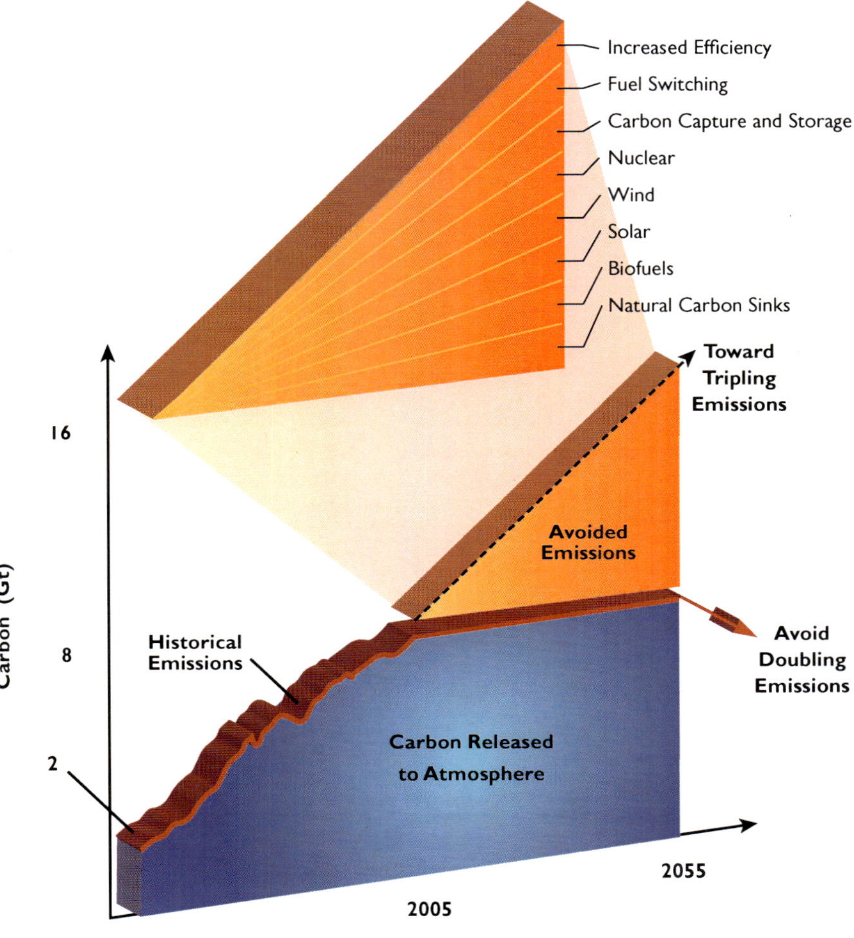

Note: Each wedge represents an increasing contribution to emissions reduction over time by that technology.

Although the concept is appealing because it makes use of proven technologies, it would not be easy to apply. For example, one wedge would come from investing heavily in new nuclear power plants to replace coal-fired plants and indirectly to replace oil used for transportation. The cost of nuclear plants is high, however, and waste storage remains a political, if not a technological, barrier. There is also concern about the potential use of spent fuel from fission to produce bombs or for terrorism.

CO_2 capture and storage could also prove problematic. Historically, to increase oil production, oil companies capture CO_2 and inject it into oil wells. In that sense, it is a proven technology, but it has not been applied at a scale necessary to capture power plant emissions, and there is little experience with burying large quantities of CO_2 over long periods of time.

Researchers have demonstrated other, newer carbon-reducing energy-production technologies in the laboratory or in pilot phases, but have not proven them at commercial scale. For example, under ideal laboratory conditions, highly efficient solar cells demonstrate as high as 40–50 percent conversion efficiency. However, producing these at commercial scale and a competitive cost poses challenges that will require considerable time and investment, and how robust such cells will prove in commercial uses is uncertain.

Many ideas regarding energy-efficiency technologies focus on motor vehicles. The Rocky Mountain Institute, for example, proposes development of much lighter and more aerodynamically efficient vehicles, made with carbon fiber composites and powered by small engines or by a hybrid combination of engines and batteries.[26] Energy-efficient building designs are possible for homes and offices, and the Rocky Mountain Institute offers architectural design tools to improve efficiencies. Radical efficiency gains in vehicles and buildings eventually would yield substantial GHG reductions, but because these types of assets are replaced slowly, the reductions would require many years to complete.

Finally, stabilization of emissions, as advocated in the wedge approach, does not imply stabilization of concentrations. As discussed in Chapter 1, enormous reductions in emissions are required to stabilize concentrations.

Issues

The investment required to mitigate climate change is in the tens of billions of dollars per year.

As noted, the federal government currently invests more than $3 billion annually in research into energy efficiency and alternate fuels. The scale of investment envisioned for a technological fix that would mitigate climate change is much greater than this current investment. The investment required is in the tens of billions of dollars per year, and it is required in production facilities as well as R&D. The capital needed for such investment could be raised by a carbon tax or by auctioning allowances under a national cap-and-trade program.

Aside from financial issues, some question remains whether government should implement particular carbon-reducing technologies or should instead conduct an extensive R&D program and leave it to private companies to make the necessary production-related investments. Some point to previous examples, such as the Synthetic Fuel Corporation initiated in the late 1970s, as an indication that government-directed investment in production facilities is not effective. A counterargument is that the scale of investment required and the need to move relatively quickly imply a more active than usual government role.

Impacts on Agencies

Federal agencies have a number of opportunities for initiating major investment programs to develop and deploy technologies to reduce the nation's carbon emissions. They can use some of the deployed technologies in their day-to-day operations. They can also incorporate these technologies into budget, maintenance, and acquisition planning.

In addition, effective deployment of some of the technologies under development may require widespread agency coordination. Carbon capture and sequestration, for example, will require environmental permitting for processing facilities and establishment of rights of way to move carbon from its production location to where it is to be sequestered. Public lands, the most convenient location for storage in some cases, are likely to be involved.

Finally, an aggressive national research, development, and deployment program will likely encompass a wide variety of technologies affecting the constituencies of several agencies. Federal leaders will have the opportunity to work with their constituencies to anticipate and adopt technologies. For example, breakthroughs in building technologies will influence the nation's housing and commercial markets, and federal leaders within agencies that deal with these industries will have the opportunity to inform constituents of ongoing technological development, assist them in the development of the necessary standards, and possibly help them secure funding to implement the technologies in their businesses.

SUMMARY

Regulatory initiatives, cap and trade, carbon taxes, voluntary programs, and technological fixes are five broad policy instruments for reducing

U.S. GHGs. Voluntary programs have been in effect for the past 2 decades and have had positive, though limited, effects, but overall emissions have continued to increase. However, as the science of climate change evolves, policymakers are taking bolder steps to reduce carbon emissions. These appear to involve a combination of regulatory initiatives, cap and trade, and technological fixes, with far-reaching implications for energy prices, technology, and the economy. Utilizing a combination of policy instruments recognizes the variety of ways carbon-based fuels are used, uncertainty about the likely success of different policy instruments, and variability in the political feasibility of those instruments.

In the short term, the budget impacts of regulatory requirements and changes in energy prices will affect agencies as new policies are established. These will affect investment choices (making fuel economy and energy efficiency higher priority investments), operating procedures, and the ability to achieve mandated missions. Over the longer term, agencies may discover opportunities, by either buying or selling carbon credits in emissions trading, or through budget support from revenues raised by auctioning emission allowances or carbon taxes. In addition, if the government engages in an aggressive technological development program, opportunities will arise to adopt some of the new technologies, work with constituents to anticipate them, and eventually adopt them as standard practice. Now is not too early to plan to take advantage of such opportunities when they arise.

---

[1] IPCC key finding. Throughout this book, each purple box shows a key finding from the IPCC with a high level of scientific understanding. See Appendix C for specific AR4 chapter sources for each finding. Chapter 1 gives a full discussion of our selection of key findings.

[2] Comparison of fleet standards among countries is difficult. Enforcement and testing varies between countries, and the U.S. standards are relatively strict. Fleet composition varies greatly between countries also.

[3] On the land-use issue, see Timothy D. Searchinger et al., "Use of U.S. Croplands for Biofuels Increases Greenhouse Gases Through Emissions from Land-Use Change," *Science*, February 29, 2008, Vol. 319. Also, Michael O'Hare, "Greenhouse Gas Emissions from Indirect Land Use Change," presentation to the California Air Resources Board Low-Carbon Fuel Standard Working Group 3, Sacramento, CA, January 17, 2008, www.arb.ca.gov/fuels/lcfs/011708UCBLUCcolor.pdf.
On the nitrous oxide (N_2O) issue, see P.J. Crutzen et al., "N_2O release from agro-biofuel production negates global warming reduction by replacing fossil fuels," *Atmospheric Chemistry and Physics,* Vol. 8, No. 2, January 2008. Debate

on these issues continues, however. For example, Wang and Haq question the sorts of land that come into production with enhanced biofuel production and how much greenhouse gas (GHG) they previously were absorbing. See letter from Michael Wang, Center for Transportation Research, Argonne National Laboratory, and Zia Haq, Office of Biomass Program, Office of Energy Efficiency and Renewable Energy, DOE, to *Science*, March 14, 2008, www.transportation.anl.gov/pdfs/letter_to_science_anldoe_03_14_08.pdf.

[4] The *Lieberman-Warner Climate Security Act of 2007* (Senate Bill 2191, introduced October 18, 2007), a cap-and-trade bill cosponsored by Senators Joe Lieberman (I-CT) and John Warner (R-VA), mandated a 5 percent cut in the carbon content of the nation's transportation fuels by 2015 and a 10 percent cut by 2020.

[5] Reed Business Information, "White Paper on Sustainability," a supplement to *Building Design & Construction*, pp. 20–21, www.bdcnetwork.com/article/CA6566572.html?nid=2886.

[6] To convert British thermal units to kilowatt hours, divide by 3412.1416.

[7] A carbon offset is a verified reduction of GHG emissions, usually a 1-metric-ton carbon dioxide (CO_2) equivalent. They must be verifiable by a third party, permanent, serialized, and tracked.

[8] *Lieberman-Warner Climate Security Act of 2007,* Senate Bill 2191, proposed by Senators Joe Lieberman (I-CT) and John Warner (R-VA), introduced October 18, 2007.

[9] GAO, *Air Pollution: Allowance Trading Offers an Opportunity to Reduce Emissions at Less Cost*, GAO/RCED-95-30, Report to the Chairman, Environment, Energy, and Natural Resources Subcommittee, Committee on Government Operations, House of Representatives (Washington, DC, December 1994), p. 37, www.gao.gov/archive/1995/rc95030.pdf.

[10] Other states and the eastern Canadian provinces have observer status. Under this agreement, states would first move toward stabilizing emissions at current levels and then seek to reduce them over a period of years.

[11] Several sources have expressed concern about the difficulties involved. See for example, Richard N Cooper, *Alternatives to Kyoto: The Case for a Carbon Tax*, Working Paper, Weatherhead Center for International Affairs, Harvard University, 2006, www.economics.harvard.edu/faculty/cooper/files/Kyoto_ct.pdf.

[12] Economists from across the political spectrum have made such arguments, including Professor Greg Mankiw, former head of the Council of Economic Advisors (CEA) under President George W. Bush, and Professor William Nordhaus of Yale University, who was a member of the CEA under President Jimmy Carter.

[13] Although the term "carbon tax" is used, usually it refers to a tax on CO_2 rather than on carbon per se. As described in Chapter 1, the two are related through a factor of 11/3, that is, a tax on CO_2 of \$Y is equivalent to a tax on carbon of \$Y × 11/3.

[14] Environmental Administration of Finland, *Environmentally related energy taxation in Finland*, October 10, 2008, www.environment.fi/default.asp?contentid=147208&lan=en.

[15] Rate as of April 1, 2008; from HM Revenue and Customs, *Climate Change Levy (CCL): Changes to rates*, BN HMRC68/07, www.hmrc.gov.uk/thelibrary/rates.htm.

[16] Congressman Pete Stark of California introduced a carbon tax bill (HR. 2069) as did Congressman John Larson (HR. 3416).

[17] Nicholas Stern, *The Economics of Climate Change: The Stern Review*, October 30, 2006, 700 pp., www.hm-treasury.gov.uk/sternreview_index.htm.

[18] The same objection can be—and is—raised in regard to cap and trade. If the United States caps carbon emissions and some other countries do not, manufacturers there receive a competitive advantage over U.S. firms.

[19] If every ton of CO_2 produced in the United States were covered under a carbon tax, a tax of \$1 per ton would generate about \$6 billion per year. Even a moderate tax, say \$20 per ton of carbon, would raise more than \$100 billion per year.

[20] Michael Canes, *A Comparison of a Cap & Trade System v. Alternative Policies to Curb U.S. Greenhouse Gases*, Washington Roundtable on Science and Public Policy, George C. Marshall Institute, February 22, 2007, p. 12, www.marshall.org/article.php?id=470.

[21] The reporting protocols of the Climate Registry can be found at www.theclimateregistry.org.

[22] EPA, *Climate Leaders*, November 19, 2008, www.epa.gov/stateply/.

[23] EPA, *SmartWay*, October 30, 2008, www.epa.gov/smartway/index.htm.

[24] William A. Pizer, Richard D. Morgenstern, and Jhih-Shyang Shih, *Evaluating Voluntary Climate Programs in the United States*, Resources for the Future, Discussion Paper 08-13, July 2008, www.rff.org/Publications/Pages/PublicationDetails.aspx?PublicationID=20485.

[25] Steven Pacala and Robert Socolow, "Stabilization Wedges: Solving the Climate Problem for the Next 50 Years with Current Technologies," *Science*, Vol. 305, pp. 968–972 (August 13, 2004), www.carbonsequestration.us/Papers-presentations/htm/Pacala-Socolow-ScienceMag-Aug2004.pdf.

[26] The concept was originated by Amory B. Lovins, Chairman and Chief Scientist of the Rocky Mountain Institute, in 1991. Lovins states that the Hypercar, as the Rocky Mountain Institute calls it, can achieve four to five times the miles per gallon of today's conventionally powered vehicles.

Chapter 4
Organizational Approaches to Managing GHGs

Offsetting the growth of or reducing emissions below current levels, as discussed in the previous chapters, will require changes in virtually every area of global energy production and use. Congress will likely initiate a national GHG management program and the executive branch may order action by executive agencies. Although a single federal agency may administer the program, agencies may have discretion in implementing the policies. This chapter presents options for implementation. It also suggests proactive steps to reduce GHG emissions ahead of regulation or beyond regulations that may emerge. For example, the National Renewable Energy Lab leads in the area of GHG reduction in government.[1]

> Both bottom-up and top-down studies indicate that there is high agreement and much evidence of substantial economic potential for the mitigation of global GHG emissions over the coming decades that could offset the projected growth of global emissions or reduce emissions below current levels.[2]

A federal leader who wants to reduce emissions will develop and implement a GHG management plan. The following sections describe GHG management plans and tools to assist in their development and implementation. They also introduce opportunities for GHG reduction at the organizational level, including improvements in buildings, appliances and equipment, transportation, waste management, and land use. The chapter closes with an assessment of the general costs and benefits of GHG management.

Although this chapter focuses on GHG management, agency efforts to *adapt* to climate change, as discussed in Chapter 6, can and should be coordinated. Chapter 6 explains how mitigation and adaptation decisions can be mutually supportive or mutually negating. Some mitigation decisions can reduce the ability to adapt and vice versa. Wise mitigation strategies provide enough time for agencies to develop successful adaptation plans.

GHG MANAGEMENT PLANS

Although multi-stakeholder GHG management and reduction programs require years of political negotiation, extensive technical capacity, and periodic revisions, the United States and other countries have already done a great deal to create such programs. The EU instituted the ETS in 2005 under the auspices of the Kyoto Protocol, and before the first trade, industry and regulators alike developed rules, standards, and technical skills.

Work done for programs like the ETS has benefited voluntary programs and regional and state regulatory initiatives in the United States, while American industry and advocacy groups have developed their own improvements in hopes of guiding regulatory requirements and spurring action. Established local, state, regional, and voluntary national programs in the United States (described below) can contribute valuable experience to the nascent comprehensive federal program. This ready-made capacity can drastically simplify efforts to address agency emissions.

The key elements of any multi-organizational GHG management or reduction scheme include

- inventory protocols,
- a reporting registry, and
- organizational management programs.

The following subsections describe completed and ongoing work in these areas and the elements of that work federal leaders can leverage.

Inventory Protocols

Many questions naturally arise when setting out to estimate GHG emissions. Frameworks developed to address these questions constitute a *protocol* for GHG reporting. A GHG reporting protocol contains two elements: requirements and recommendations. In the most widely used and accepted protocol in the United States, these are called standard and guidance:

- The standard is the core set of practices agencies must follow to comply (Table 4-1).
- Guidance sections explain choices available in the standard, illustrate their implications, and provide examples of implementation.

Table 4-1. Four Standards of GHG Protocol

Standard	Details and decisions
General accounting principles	Relevance, completeness, consistency, transparency, and accuracy
Organizational boundaries	Equity share or control approach
Operational boundaries	Identify emissions related to operations; categorize them by scope
Reporting emissions	Use best available data, report changes, and treat credits separately

The general accounting principles of a GHG protocol conform to those in any accounting situation. The reporting agency should report only the information needed and should render all the information needed to understand the report. Reports must be consistent from one application to another and from one period to another. Agencies must explain any deviations or changes, and all underlying data must be available. The reporting body must maintain a level of accuracy that is well documented and appropriate to the situation.

The GHG Protocol offers a standard method for GHG reporting that is widely accepted. It also offers tools and supporting documents (ghgprotocol.org). International Organization for Standardization 14064 Part 1 is based on and fully compatible with the GHG protocol corporate standard.

In considering organizational boundaries, most federal agencies will prefer to report emissions for activities under their control. The concept of equity share is relevant in few situations in the federal realm.

Operational boundaries establish the GHG emissions[3] related to the operations of the agency. They are from sources owned or controlled by the agency, such as boilers, vehicles, or furnaces (Scope 1); from the generation of purchased electricity consumed by the organization (Scope 2); or a consequence of the activities of the agency but from sources not owned or controlled by the agency (Scope 3). Some examples of Scope 3 activities are transportation of purchased fuels, constituent compliance with policies developed by the public-sector organization, or the fuel consumed by visits to a national park. Figure 4-1 illustrates these three scopes.

Figure 4-1. GHG Scopes

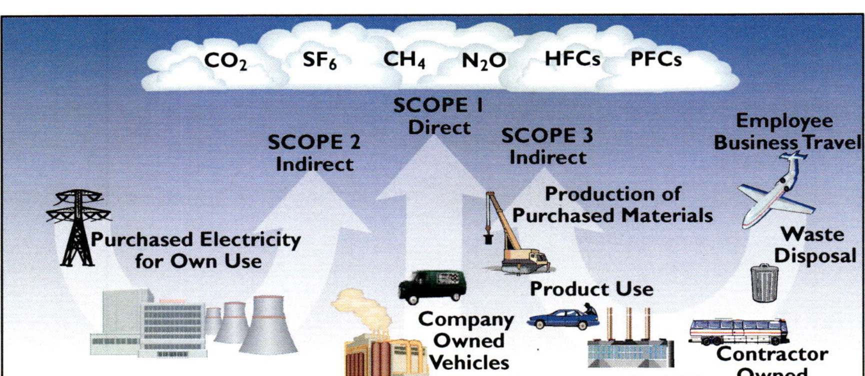

Note: SF_6 = sulfur hexafluoride.

The GHG Protocol is the basis for nearly all reporting programs in the United States,[4] such as those discussed below.

Registries and Reporting Programs

Two key federal reporting programs shape U.S. GHG management. Section 1605(b) of EPAct 1992 established one of the first mechanisms through which anyone in the United States could voluntarily report GHG emissions. The EPA Climate Leaders program is another federally administered voluntary reporting program, which uses a partnership-based approach to engage corporations. Climate Leaders goes further than 1605(b), using internationally recognized inventorying protocols and helping its partners develop inventories that are more robust and work toward GHG reduction goals.

Because there has been little centralized and coordinated effort to develop an overarching set of requirements and standards for GHG reporting and management beyond these two relatively small federal programs,[5] multiple and sometimes disparate industry, state, and local government programs have proliferated. The state of California established the California Climate Action Registry (CCAR) in 2001 at the request of industry leaders who wanted to ensure that they would receive credit in future regulatory schemes for their early responses to climate change. CCAR served as the basis for The Climate Registry,[6] which is now the largest reporting registry for GHG emissions in North America—covering 39 states and 3 tribes in the United States, 6 states in Mexico, and 9 provinces in Canada—with 289 members as of November 11, 2008. Many of those organizations report as members of the WCI.

Reporting registries serve as clearinghouses for both voluntary and mandatory organizational GHG inventories. They develop unifying rules and procedures so that individual reporting entities that may fall under a range of different reporting programs can all report to one body. They aim to simplify reporting processes while providing a central point of access for data and best practices for users and the public.

Various reporting programs have standalone registries, such as the CCX, while others, such as one created by California's Global Warming Solutions Act of 2006 (Assembly Bill 32), are in the process of determining reporting requirements. For its part, The Climate Registry receives inventories from an array of voluntary and mandatory reporting programs and entities. It records and tracks the inventories reported by northeastern U.S. energy producers regulated under the RGGI cap-and-trade scheme.[7] The Climate Registry is also working with the WCI and International Council for Local Environmental Initiatives (ICLEI)—Local Governments for Sustainability to develop their reporting requirements.[8] More than 1,000 local governments now belong to ICLEI.

Although the array of voluntary and mandatory reporting programs with varying goals, covering a range of geographical areas and reporting entities, appears fragmented, the connections they have to overarching registries unify them. The result is readily accessible tools and lessons learned from experienced GHG management programs. Not only are these resources useful for federal leaders, as federal-level involvement increases, but many may become the standard. For example, the Lieberman-Warner Climate Security Act identifies The Climate Registry as a model system for a national-level reporting program.

GHG Management Programs

> There is high agreement and much evidence that: Decision making about the appropriate level of global mitigation over time involves an iterative risk management process that includes mitigation and adaptation, taking into account actual and avoided climate change damages, co-benefits, sustainability, equity, and attitudes to risk. Choices about the scale and timing of GHG mitigation involve balancing the economic costs of more rapid emission reductions now against the corresponding medium-term and long-term climate risks of delay.[2]

For any organization to efficiently and effectively mitigate the risks and limit the impacts associated with its GHG emissions, it must develop a comprehensive management program. GHG management programs provide a clear path to meeting reduction goals by

- establishing a responsible and accountable team,
- developing a complete inventory of amounts and sources of emissions, and
- identifying the most reasonable mitigation approaches for an agency's given circumstances.

Federal agencies operate under unique conditions. Most obviously, the mission of an agency and its subordinate operating units pervades every decision made and action taken. Political and economic realities affect budgets, staff sizes, and personnel capabilities. Geographic dispersal across states and countries can dictate the laws that apply to different parts of the organization and thus can create an array of compliance requirements. These and other agency-specific nuances influence how agencies should react to GHG-related risks and opportunities and thus should inform how an agency manages its emissions.

An agency-wide program will set high-level policies, disseminate information, and coordinate inventory consolidation for reporting to the designated registry. Subordinate-level management programs will be necessary to deal with issues unique to specific locations or types of operations. The various levels of GHG management programs must be coordinated to ensure consistency throughout the organization and to prevent any activities from being overlooked.

Management plan implementation is a continuous process. Changes in external and internal influences on program execution, such as regulatory requirements, personnel, technology improvements, and budget availability mean that the program will require continuous updates and improvements. Table 4-2 outlines the basic steps and elements involved in a GHG management program; note that the last step requires cycling back to the beginning to reevaluate and update program elements as necessary.

Table 4-2 highlights the key steps that a federal leader should follow in the development and execution of a GHG management program. The few federal agencies that have begun to implement a program have highlighted the importance of taking advantage of data and expertise that already feed into other reporting requirements. In particular, agency energy managers collect data for energy consumption reports to FEMP, the same data that GHG emissions reporting will require.

Table 4-2. Steps for Developing a GHG Management Program

1.	**Assign responsibility and organizational structure to facilitate program implementation and management.**
	Determine elements involved and leader. Develop a cross-functional executive steering committee to align efforts. Consider related responsibilities (such as energy consumption reporting) and available resources. Align with legislated authority (for example, the agency environmental executive) and the agency's mission.
2.	**Evaluate risks and opportunities.**
	Risks may include being ill prepared for regulation and reputational risk. Opportunities may include funding for specific mitigation options or enhanced multipurpose data management systems, setting the example, and influencing suppliers.
3.	**Set program goals and objectives.**
	These may include carbon neutrality, voluntary reductions, leadership, cost savings, or compliance.
4.	**Develop a GHG accounting system; conduct a baseline emissions inventory on which to base evaluation of program progress.**
	Institutionalize procedures to ensure high quality. Piggyback other requirements. Create a structure that will allow for easy rollup of disaggregated data. Involve and develop in-house expertise.
5.	**Identify and prioritize options for achieving program goals.**
	Rank options on the basis of organizational mission, costs, and ability to reduce emissions. Analyze costs and benefits of options; determine applicability of offsets and credits.
6.	**Deploy mitigation and reduction techniques.**
	See "Portfolio of Practical Solutions" section.
7.	**Measure and report progress.**
	Recognize multiple reporting requirements (international, federal, state, and regional).
8.	**Review and revise.**
	Have senior management review each step. Implement refinements.

Taking advantage of the systems and data quality assurance measures already in place will maximize the use of resources. For example, many agencies have energy management systems, and others operate sustainability management systems. Integrating GHG management programs into existing environmental and energy management programs and requirements may be prudent, and some agencies and federal facilities have already done so. This integration is easier if the GHG team includes representatives from different traditional organizations (energy, transportation, environment, etc.).

PORTFOLIO OF PRACTICAL SOLUTIONS

A GHG management plan provides a baseline for reducing GHG emissions, and the IPCC suggests that many of the reduction options available promise to help achieve stabilization levels as discussed in Chapter 2.

> There is high agreement and much evidence that all stabilization levels assessed can be achieved by deployment of a portfolio of technologies that are either currently available or expected to be commercialized in coming decades, assuming appropriate and effective incentives are in place for development, acquisition, deployment, and diffusion of technologies and addressing related barriers.[2]

No single approach known or likely to emerge offers the needed reduction now or in a practical period. Combining an array of technologies and developing policies to support their expansion in the market can achieve stabilization (see Chapter 3). Energy efficiency and increases in low-carbon energy sources are key focal areas for reducing GHGs—but insufficient in themselves. A full portfolio of methods must be applied, and this chapter describes those directly available to the agency leader.

Large-scale projects, such as installing megawatt-scale wind farms or retrofitting all facilities to meet green building standards, may be beyond the budgets of most federal leaders. However, a range of technological and process improvement options are available for reducing GHG emissions on a smaller scale, which organizational leadership can often direct or champion. For example, making relatively small adjustments to water and room temperatures can reduce fuel consumption significantly. In the United Kingdom, a reduction of just 1˚C cut fuel consumption in a building by 8 percent.[9] What other options and tools are already available?

A few goals capture the intentions of virtually all GHG reduction options:

- Minimize energy waste.
- Use energy more efficiently.
- Replace fossil-based energy with renewable sources.

- Sequester GHGs.
- Influence emissions of suppliers and constituents.

Some of these goals are incorporated in federal requirements such as EPAct 2005 (see Chapter 3) and EO 13423 (see Chapter 1). Many require changes in operations or the behavior of personnel. Achieving lifestyle changes may require an enormous effort and market change. Education and training programs can help overcome barriers to market acceptance of energy efficiency.

The concept that the cheapest energy is that which is not used (the so-called *negawatt*) applies to both operational costs and GHG emissions. Simply powering down unused office equipment, turning off office lights at night, turning off idling vehicles, or switching from a paper-based to an electronic-based filing system can reduce energy use and GHGs without necessarily requiring new technologies. Many of these energy reductions can be automated through use of timers or occupancy sensors.

The federal government manages substantial numbers of people, programs, projects, products, and services and thus is responsible for significant carbon flows. The government employs more than 2.7 million people (2 percent of the American workforce) and hires approximately 300,000 employees each year.[10] Although the amount a single individual could reduce GHG emissions may be relatively small, initiatives that encourage all federal employees to make GHG reductions could have a significant cumulative impact on the nation's GHG emissions (see "Waste Management" below).

Programs to encourage employee behavior changes or modify operational processes can be inexpensive and easy to merge with training or other improvement initiatives. For example, FEMP's "You Have the Power" campaign,[11] begun in 1997, helps federal energy managers promote energy efficiency. The program offers training programs and educational handouts, posters, special publications, and other targeted outreach materials.

EPA currently runs a variety of programs that encourage behavioral change, such as its

ENERGY STAR Challenge, which asks all Americans to make small changes in their daily lives that will result in GHG reduction.[12] These efforts complement federal energy goals and legislative drivers.

Increases in energy efficiency through technology and material upgrades can decrease GHG emissions by reducing demand for energy at the power plant. Efficiency-improving technologies allow the same amount of productivity but require less energy input. Common examples are compact fluorescent light bulbs and diesel vehicle engines, which allow the same amount of light and the same distance of travel, respectively, using less source energy.

Upgrading to higher-efficiency equipment generally entails increased up-front costs, but over the life of the equipment, the reduced energy costs (and often increased lifespan) can result in an overall positive net return on investment. The energy savings lessen the pressure on agency budgets while reducing GHG emissions.

More than 78 percent of U.S. GHG emissions are from energy production from fossil fuels (particularly coal and petroleum).[13] Cutting the use of carbon-based energy offers great potential for GHG emissions reduction. A recent analysis (detailed in the final section of this chapter) ranked 250 GHG abatement options according to cost and ability to reduce GHG emissions.[14] The most economical is eliminating energy waste. Federal agencies consume much energy and could greatly reduce waste in three areas: transportation, buildings, and appliances and equipment. We discuss each of these below.

Agency Scope 1 Emissions

Reduction in transportation fuel use is one of the most discussed ways to reduce energy use in the federal government. Transportation emissions accounted for 33 percent of U.S. CO_2 emissions in 2005. In addition to these individual contributions, the fleet of federally operated government vehicles produced roughly 3.5 million tons of CO_2 equivalent in 2007.[15] Civilian agencies operate more than 230,000 vehicles, and DoD alone operates nearly 193,000 nontactical and tactical vehicles. The USPS has a fleet of nearly 218,000 vehicles, also a significant source of GHGs.

Federal leaders can influence policies and programs involving agency fleet vehicles. Consolidating trips and passengers can reduce petroleum consumption in federal fleets, yet work circumstances can limit this opportunity. Procurement of alternative fuel or fuel-efficient fleet vehicles is more costly, but statutes such as EPACT 2005 and EO 13423 support this type of procurement. Downsizing vehicle power capacities to meet requirements can improve fuel efficiency. In addition, purchasing decisions should factor in the tradeoffs among utility, cost, and fuel efficiency found in a range of engine and drive train types, such as gasoline, diesel, and hybrid-electric vehicles.

A federal leader can plan for future procurement and leasing of vehicles, demanding fuel-efficient vehicles for the agency's fleet. Table 4-3 contains "well-to-wheels" data that show the relationship between miles per gallon and carbon equivalent per gallon for a variety of vehicles.[16] Well-to-wheels analysis includes the GHG emissions associated with getting fuel to the consumer as well as use in the vehicle itself. For a gas-powered auto, it is equivalent to multiplying the mpg by a factor of 0.83. To put Table 4-3 in more familiar terms, the emissions of a single gas-powered auto for a typical year of use (12,000 miles) are more than 7 tons of CO_2.

Table 4-3. Comparison of Fuel Mileage and GHG Emissions

	GGE	CO_2-eq (grams/mile)
Gasoline ICE	21.3	552
Diesel ICE	25.8	442
Gas-electric hybrid	29.2	406
E85	21.3	451
E85 (cellulosic)	21.3	154
CNG	21.0	438
Public transit		
Heavy rail	136.7[a]	156[b]
Commuter rail	75.5[a]	177[b]
Bus	28.6[a]	56[b]
Van pool	89.7[a]	101[b]

Note: GGE = gallons of gasoline equivalent; ICE = internal combustion engine; E85 = Ethanol 85 percent; CNG = compressed natural gas.

[a] GGE per passenger mile.

[b] CO_2 grams per passenger mile.

Biofuels such as ethanol and biodiesel have garnered much attention as a means to reduce GHG emissions. As with other biobased products, the claim of reduced GHGs relies on the fact that crops such as corn and soybeans take in CO_2 from the atmosphere during their growth cycles and release it during fuel combustion, thus "recycling" the CO_2 from one growing season to the next. However, fertilizers, fuel processing, and fuel transportation all involve CO_2 that is not "recycled" during crop growth. Furthermore, some studies suggest that increased production of biofuels has resulted in deforestation to accommodate crop production, resulting in the loss of carbon-storing trees. In addition, using food crops to fuel vehicles has sparked concerns over food prices and food availability around the globe. The debate concerning the true value of biofuels will continue as long as these issues remain, but "second generation" biofuels, such as cellulosic ethanol and algae-based biodiesel, offer hope that biofuels can play a significant part in reducing GHG emissions without requiring difficult tradeoffs.

A lack of fueling infrastructure and engine compatibility limits the expanded use of biofuels, just as it did with natural gas in the past. Fueling infrastructure issues may not be as significant for fleet vehicles as nonfleet vehicles because their limited radii of operations allow for the use of fewer refueling stations. For most alternative fuels, engine compatibility can only be addressed through the purchase of new vehicles or engine modifications.

The first opportunity for GHG reduction focuses on energy reduction—using less energy usually results in less GHGs. However, this is true only when one considers energy reduction over the life cycle of a product, so simply looking at tailpipe emissions reveals only part of the story. For example, replacing petroleum-powered vehicles with hydrogen fuel cell vehicles will result in reduced oil consumption and fewer GHGs emitted directly from vehicle operations, but may increase overall energy requirements and GHG emissions.[17] The entire production chain of petroleum fuels may involve less coal-based electricity than that of hydrogen. In other words, using less petroleum does not necessarily mean reduced energy needs and less GHG emissions.

Agency Scope 2 Emissions

Buildings

> There is high agreement and much evidence that: Energy efficiency options for new and existing buildings could considerably reduce CO_2 emissions with net economic benefit. Many barriers exist against tapping this potential, but there are also large co-benefits.[2]

The nonresidential end-use sector accounted for 18 percent of CO_2 emissions from fossil fuel combustion in 2005. Seventy-eight percent of these emissions were from electricity consumption for lighting, heating, cooling, and operating appliances. More than 9 percent of the energy produced in America's power plants goes to operate nonresidential buildings,[18] and EPA estimates that 35 percent or more of that electricity is wasted through inefficient building operation and management strategies.[19]

Although the majority of this building space belongs to the private sector, the federal government manages more property than any other single entity in the United States. CO_2 is emitted as a result of the operation of each of these buildings, and reductions in fossil energy use in federally owned and operated buildings could significantly reduce GHG emissions. Improving energy efficiency in buildings and appliances could result in the abatement of 710–870 million tons of CO_2-eq at relatively low cost.[14] This is several percent of the total GHG emissions in the United States (Figure 1-2).

The General Services Administration (GSA) owns and leases more than 340 million square feet of space comprising 8,920 buildings in more than 2,000 communities nationwide. In addition to these administrative buildings, it also manages courthouses, border stations, laboratories, post offices, and data processing centers across the country. DoD also owns thousands of administrative and residential buildings in the United States, as well as outside the country.

DOE is working on several fronts to encourage and assist with energy-saving building design. For example, the Energy Plus modeling program, which is available for free download, models energy flows and includes innovative variables such as natural ventilation and photovoltaic systems.[20]

LEED (see Chapter 3) offers a useful overarching system for demonstrating that buildings are constructed and operated at the highest available level of energy efficiency. Several federal agencies, including DoD, USDA, DOE, and DOS, have LEED initiatives. As of 2003, all new GSA building projects are certified through the Leadership LEED Green Building Rating System, and a Silver LEED rating is encouraged.

Lighting accounts for the largest single portion of energy consumption in buildings, followed by space heating and cooling (Figure 4-2). Efficient lighting and day-lighting; efficient electrical appliances; passive and active solar design for heating and cooling; heating, ventilation, and air conditioning retrofits; alternative refrigeration fluids; and recovery and recycling of fluorinated gases all can reduce GHG emissions,[21] and these technologies are currently available commercially. The most cost-effective targets for buildings are appliance efficiency, refrigerators and freezers, ventilators and air conditioners, water-heating equipment, and lighting best practices.[21]

Figure 4-2. U.S. Commercial Building Energy Use in 2005

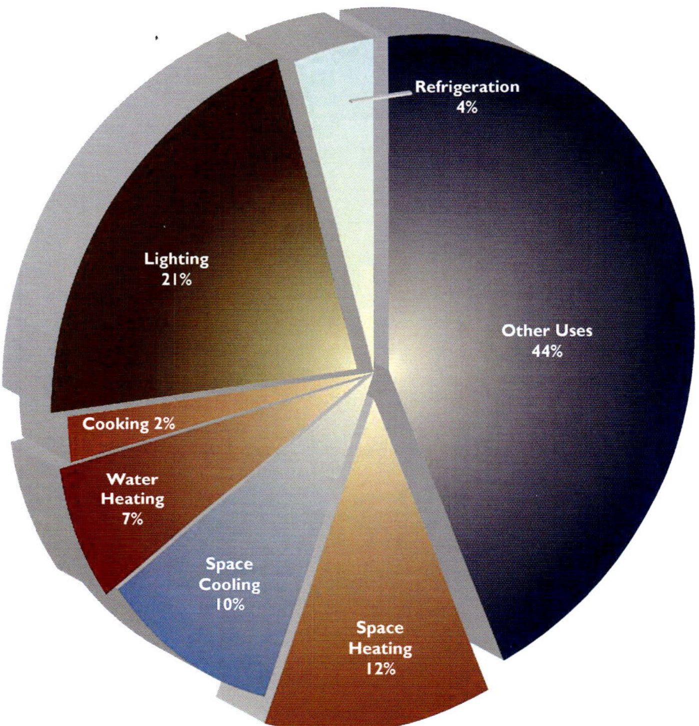

Changes in occupant behavior and choices and use of technologies can considerably reduce emissions related to energy use in buildings. Employee actions as simple as turning lights off when rooms are not in use (a switch-off policy), modifying room temperature, or using stairs instead of elevators can reduce a building's energy requirements. Managers may find it challenging to achieve changes in employee behavior. Understandably, employees may not respond well to efforts to mandate certain actions. However, simple efforts to instill an energy-aware organizational culture can be as simple as adding an energy-efficiency training session to annual training requirements. Occupancy sensors to control lights can reduce lighting costs by 30 percent. Table 4-4 shows options for replacing traditional light bulbs with more efficient compact fluorescent light (CFL) bulbs.

Table 4-4. Light Bulb Replacement

Standard incandescent (watts)	CFL replacement (watts)	Load savings (%)
25	7–9	64–72
40	11–13	67–73
60	15	75
75	18–20	73–81
100	28–30	70–72

EPA's Green Lights Program encourages partners to shift their thinking about lighting—from an obligatory part of the building to an opportunity to save money and help the environment. EPA offers the opportunity to sign the Change a Light Pledge on its website, and federal organizations have taken on the challenge by signing the pledge. DoD leads in this program, having more than 9,000 individual pledges and saving more than 11 million kWh as a result of the bulbs replaced.[21] DoD's success may be attributable to the 2008 ENERGY STAR Operation Change Out Military Challenge. All service members are being encouraged to save energy and money and protect the environment by replacing their inefficient, incandescent light bulbs with ENERGY STAR–qualified bulbs.

Building management offers another variety of solutions for reducing GHGs. The DOE FEMP site provides ample guidance on selection of energy-efficient materials for buildings.[22] The Carbon Trust presents a comprehensive list of ways to reduce energy use through behavior changes in its guide for local authorities (Table 4-5).

As mentioned, GSA owns and operates a large fraction of federal buildings. The arrangements of ownership are many and varied. For example, GSA may lease buildings from private companies and sublet them to other federal clients. Depending on the leasing agreements, GSA may have no control over utility use in the building. To complicate matters, nonfederal clients may lease space in the same buildings. In addition, GSA owns many buildings for which it controls electricity supply, but in which electricity demand is driven by occupants' activities. In many cases, electricity is monitored only for the entire building and not for individual occupants. In short, the variety of arrangements of federal ownership and control introduce significant challenges in GHG reporting and mitigation and complicate motivating efficient use of buildings. Agencies that don't own a building may not be motivated to conserve.

Table 4-5. Opportunities for Energy Consumption Savings

Building operation	Energy-saving practice	Description
Heating and hot water	Keep the heat in	Avoid opening doors and windows when heating is in operation
Heating and hot water	Get feedback to take action	Staff should report areas that are hot, cold, or drafty
Heating and hot water	Consider the season	Consider the attire that staff will be wearing in certain seasons
Heating maintenance	Keep systems clear and unobstructed	Keep furniture away from vents
Heating maintenance	Maintain boilers and pipework	Have equipment serviced regularly
Heating maintenance	Control	Match system control and timing to particular facility
Heating maintenance	Consider building energy management systems	Use a network of controllers to optimize energy use
Hot water	Check hot water temperature	Make sure water is not too hot to keep hands under tap
Lighting	Switch-off policy	Turn off lights when leaving a room
Lighting	Avoid blinds down and lights on	Encourage staff to direct blinds to ceiling and walls
Lighting	Label light switches	Help staff know which switch controls which light to avoid overuse
Lighting	Install low-energy lighting	Replace traditional lighting with low energy options and sensors
Ventilation and air conditioning	Naturally cool for free	Do not use air conditioning when external temperatures are comfortable
Ventilation and air conditioning	Night cooling	Use naturally lower temperatures at night to cool down the building

APPLIANCES AND EQUIPMENT

Like buildings, appliances and electronic equipment represent an opportunity to reduce energy consumption. Turn off appliances or turn them down. Power down computers at the end of the day. Set refrigerators to avoid excessively cold temperatures. In procurement, ensure that the size and functionality of appliances match the needs.

However, the behavioral change approach can only go so far. Employees still need to turn on lights and computers and will continue to demand cold drinks. This calls for continuous improvement in the energy efficiency of the items used most frequently. A vast array of relatively inexpensive energy-efficient appliances is widely available. The ENERGY STAR program offers comprehensive information on energy-efficient appliances. Its website tools identify ways to increase energy efficiency in office appliances.[23]

Federal organizations can realize reductions in the same way that the private sector does. EPAct 2005 requires federal agencies to purchase ENERGY STAR or FEMP-qualified products in the top 25 percent of their class for energy efficiency.[24] In addition, EO 13423 (introduced in Chapter 1) requires 95 percent of federal agency electronics purchases to meet the Green Electronics Council's Electronic Product Environmental Assessment Tool standards. The order also requires federal agencies to enable the ENERGY STAR "sleep" mode on agency computers.

Agency Scope 3 Emissions

While Scope 1 and Scope 2 GHG emissions are the most likely to be regulated in the midterm, Scope 3 emissions can be an optional part of an agency's emissions reduction plan. Here we present aspects of agency operations that are now influenced by existing regulations and that involve significant opportunities to reduce Scope 3 GHG emissions, for example, meeting green and affirmative procurement goals. EISA already specifies procurement of alternative energy and requires a life-cycle GHG analysis, but there is not yet an accepted standard for that analysis. Approaches to life-cycle GHG analysis are discussed further below.

Scope 3 emissions are difficult to estimate with current data, so the federal leader must balance a desire to reduce GHG emissions with the cost and effort as well as the lack of a mandate for a comprehensive program of mitigation.

Employee Commuting

The federal government employs more than 1.8 million civilian employees, excluding those in the USPS.[25] A large percentage of these individuals drive to and from work each day, personally contributing to GHG production. Although these emissions are not directly attributable to the federal government, they represent a significant source associated with the federal workforce and therefore an opportunity for reductions through telecommuting practices.

Telecommuting has proven successful for GSA, which implemented a pilot program in 2005. As reported in May 2007, 113 federal employees participated in the program, working 2 days per week on average from a telework center or their home. GSA estimates that its telework program results in the reduction of 2.3 million pounds of emissions annually.[26] The number of executive branch employees who worked at least 1 day per month from home was 110,592 in 2006, or 6.1 percent of the executive branch workforce.

Federal agency leaders encourage employees to use public or non-motorized transit or provide information and education aimed at reducing automobile usage and improving efficient driving techniques. Other federal programs can be counterproductive in reducing GHG emissions. For example, parking subsidies encourage commuting. Balancing the goals of these efforts (employee benefits, quality of life, traffic reduction, and GHG mitigation) creates tension in personnel management.

Waste Management

> There is high agreement and much evidence that post-consumer waste is a small contributor to global GHG emissions (<5%), but the waste sector can positively contribute to GHG mitigation at low cost and promote sustainable development.[2]

Improvements in waste disposal practices, reduction of waste, and increased recycling can reduce GHG emissions (as noted in the IPCC finding above).[27] As key mitigation practices, existing low-cost waste management options (recycling, waste minimization, landfill methane recovery, waste incineration with energy recovery, and composting of organic material) result in emissions mitigation and improved public health and safety.[21]

EPA has been working to manage waste for decades. For example, in 2002, it launched its Resource Conservation Challenge, urging Americans to meet or beat two goals by 2005: boosting the national recycling rate from 30 percent to at least 35 percent and curbing the generation of 30 harmful chemicals normally found in hazardous waste by 50 percent.[28] EPA may play a significant role in helping the nation reduce GHG emissions through waste management.

Reducing waste in federal government office products will reduce energy use and GHG emissions. As Table 4-6 shows, the typical office that switches from the base case (non-recycled paper) to paper with 30 percent recycled material saves about half a short ton (948 lb) of carbon emissions each year, and switching to 100 percent recycled paper saves more than 1½ short tons of carbon emissions.

The typical office that switches from non-recycled paper to paper with 30 percent recycled material saves about half a short ton of carbon emissions each year.

EPA's WasteWise program, a voluntary effort aimed at eliminating municipal solid waste and select industrial wastes, has more than 2,000 partners,[29] including 339 from the federal government. Joining this program can help reduce agency GHG emissions. The Climate Profile tool converts waste reduction data provided by partners to tons of carbon equivalent reduced. In 2005, the partners eliminated more than 8 million metric tons of carbon-equivalent emissions.

Table 4-6. Life-Cycle Carbon Emissions from Office Paper

Paper used	Life-cycle lb CO_2-eq per short ton of paper	Total carbon (lb)	Carbon savings from base case (lb)
0% recycled (base case)	5,690	8,535	—
Current EPA CPG standard (30% recycled)	5,058	7,587	948
100% recycled	3,582	5,373	3,162

Note: The typical U.S. office worker uses about 10,000 sheets (100 lb) of copy paper each year. A 20-person office would use 1 short ton of paper per year.

CPG = *Comprehensive Procurement Guideline.*

The New York City GHG inventory found that the overwhelming percentage of waste generated by office workers in the city consisted of paper.[30] The federal government has already taken strides to reduce excess paper use and to purchase environmentally friendly supplies. In addition to following EO 13101, numerous agencies have gone beyond its requirements to reduce paper waste. For example, EPA developed a purchasing agreement that allows all its offices to order cost-competitive paper with 100 percent recycled content. EPA Region 10 began using "Good on One Side" (GOOS) in 2006, encouraging the staff to reuse paper for informal purposes. EPA Region 10 reduced paper consumption by 30 percent after implementing the GOOS paper program and changing its buying practices.[31] In addition, the Federal Network for Sustainability offers the Greening Federal Copier Paper Wizard,[32] which furnishes essential information to help federal agencies transition to greener paper practices and purchasing.

GAO released a report in 2005, *Electronic Waste: Strengthening the Role of the Federal Government in Encouraging Recycling and Reuse*, which highlights the need for improved purchasing, operating, and disposal practices for electronics in the federal government.[33] The report explained that the primary concerns over electronics are a waste of natural resources and disposal of hazardous substances. The Federal Electronics Challenge seeks to tackle these problems by encouraging federal facilities and agencies to purchase greener electronic products, reduce impacts of electronic products during use, and manage obsolete electronics in an environmentally safe manner.[34] Originally, this challenge did not focus on climate change, yet it is relevant to the subject now, as reducing waste of products can reduce an organization's carbon emissions. Many of these GHG reductions are Scope 3 and may be outside of emission regulations likely to be implemented in the near term.

Upstream and Downstream Emissions

As noted in the discussion of operational boundaries above, federal agencies influence GHG emissions from activities beyond their control. Specifically, contractor activities and constituent responses to agency policies may not fall directly into agency operations, but they exist because of an agency action or decision. Although generally considered optional for reporting purposes, these types of emissions may represent the most significant volume that an agency leader can influence. Below are some opportunities for emissions reductions beyond agency activities.

Contractors do a vast amount of work at the behest of the federal government in activities from food preparation to weapons system production. In many cases, contractors operate in their own facilities, deciding which equipment or processes to use as they work under a federal contract. However, federal leaders can greatly influence the structure of contracts and statements of work and thus may encourage or require contractors to reduce GHGs or to use low-carbon techniques in their work.

Organizations have already begun to include related environmental requirements in contracts. For example, Fairchild Air Force Base developed a "green statement of work" for its custodial services.[35] Federal managers can look to stipulate these types of requirements in the contracts they develop and can even use the contracts to develop the capacity to track the GHG emissions of their contractors.

Policy decisions are another agency activity that can influence downstream GHG emissions. In general, when an agency analyzes the potential impacts of a policy it is considering, it thoroughly analyzes costs and benefits; this is particularly true for policies that may have a significant environmental impact. However, decision makers often overlook the environmental impact of policy decisions. Including those implications in the process may have beneficial results.

For example, the No Child Left Behind Act of 2001 allows students to transfer to an alternative school if their school does not meet required standards in a set period. If this results in the physical transfer of students to a new location, it could result in increased fuel use by the associated school district. Conversely, if the act results in an improvement in the public schools, some constituents who previously transported their children to private schools will reenroll them in their neighborhood public schools, reducing transport-related GHGs. These considerations are unrelated to goals of the law, but they will affect overall GHG emissions and may impact a school district's ability to meet its GHG goals.

Another example stems from the post-disaster cleanup of debris and housing materials scattered by Hurricane Katrina in 2005. To rebuild New Orleans and the surrounding area, crews cleared large volumes of materials to make way for new construction. However, concerns over landfill capacity and propagation of the aggressive and invasive Formosa termite led recovery planners to investigate waste management options other than landfill disposal. The two leading candidates are on-site combustion and mechanical grinding (to reduce volume). Because each option releases a range of harmful particles and pollutants, EPA modeled the impacts of large-scale implementation to determine which is least harmful to human health and the environment. The analysis covered 65 pollutants, but not the operations' GHG emissions. Although GHG impact may not or should not be the primary criteria for this decision, EPA could factor it in as part of a more robust decision. Further, GHG concerns could spur the development of a modified solution, such as adding energy recovery to the combustion option.

Green Supply Chain

OMB estimates that in 2009 the federal government will spend more than $180 billion on supplies, materials, and equipment.[36] All of these items have an associated carbon emissions from the production and distribution of the material. These emissions are an indirect responsibility of the federal government in that federal leaders can use GHG footprint as an evaluation factor in selecting suppliers. Currently, selection factors do not typically include consideration of GHG footprints, primarily because there is no standard calculation method.

In addition, the federal government—especially DoD, GSA, DOS, and the Department of Homeland Security (DHS)—operate significant supply chain activities, which have their own GHG emissions such as for the building and fleet emissions mentioned earlier. However, these supply chains also directly contract for building, transportation, maintenance, and other activities. Although the government does not own these activities, it is responsible for their emissions.

These procurements and activities represent multiple supply chains throughout federal operations. The sheer volume of activity combined with the multiple organizations (governmental and commercial) involved in a typical supply chain make managing the emissions from the supply chain a challenging task. However, a structured approach will help the federal leader manage the emissions from the supply chain.

Managing GHG emissions in the supply chain relies on two primary concepts: first, working with suppliers, contractors, and, potentially, agency customers to identify and reduce emissions; second, optimizing across activities to reduce overall GHG emissions from the supply chain.

To be able to manage the emissions from external organizations (suppliers, contracted operations, etc.), the federal leader must reach out to those organizations. In general, leaders should approach this outreach as a partnership to work together to identify GHG sources, measure emissions, and undertake programs to control and reduce them. The federal leader also can include GHG management programs in contract statements of work and as an evaluation factor in the procurement process. To do so, standards and metrics are needed to evaluate success. The U.S. Air Force has faced this challenge in developing bio-based jet fuel. Working with supply chain partners will help identify ways to reduce both direct and indirect emissions.

Beneath the complexity of the supply chains, all are a process for procuring material and delivering it to the point of use. Looking at the supply chain as a process can help avoid suboptimizing GHG emissions. Suboptimization occurs when emissions from one activity or location are reduced, yet total supply chain emissions increase. Evaluating the supply chain as a single process identifies total supply chain emissions and allows informed decisions on the best activities to address and the best approach for reducing GHG emissions.

Renewable Energy and Carbon Sequestration

Renewable Energy Sources

EO 13423 already mandates increased use of new sources of renewable energy by federal agencies (see Chapter 1). To implement this EO and other regulations that Congress may impose, the federal leader must understand potential sources and new developments in the field. Below, we review those sources and developments.

Renewable energy technologies, such as solar photovoltaic cells, wind turbines, geothermal heat pumps, and ocean-based energy, can displace fossil fuels with clean, emissions-free energy sources. Renewable energy infrastructure can range from small-scale, distributed units that generate power for individual facilities or equipment to large-scale "farms" that can power entire towns. Renewables can provide energy to meet "off-grid" power demands, which is a key advantage for remote facilities constructed far from electrical distribution infrastructure or when considering backup power supplies to reduce the risk of grid-related failures. What's more, most states provide for net metering capabilities that allow for the sale of excess power generated from distributed renewable sources "back to the grid," creating a potential source of income.[37]

Three of the top 25 renewable energy purchasing organizations in the United States are federal government agencies. EPA and DOE are in that group, but the leading federal organization is the U.S. Air Force, coming in at number 3 in the EPA's 2008 listing of top 25 green power purchasers. These organizations are setting an example for all other federal organizations. The U.S. Air Force is currently purchasing 9 percent of its power from green sources.

GSA works to increase renewable energy use across its facilities and purchased almost 950 million kWh of energy from renewable sources in 2003–07. Examples of renewable energy include the following:

- Binghamton Federal Building in New York state, the first federal facility powered by 100 percent renewable (wind turbine) energy
- Statue of Liberty and Ellis Island, with electricity generated from 100 percent wind resources
- Denver Federal Center Solar Park, which generates about 1.6 million kWh of energy each year from an array of solar panes.

GSA estimates that 543,700 kWh of the total renewable energy in FY06 came from 12 solar photovoltaic installations, 175,800 kWh came from GSA's two solar thermal projects, and 243,200 kWh came from the one completed geothermal project.[26] These examples reduce emissions and give the organization prestige as a leader in the use of renewable fuels.

With increases in market penetration and investment in renewable energy sources, improvements in "conventional" renewables such as solar photovoltaics and wind will make them even more appealing. For example, new materials for wind turbines and solar cells will improve efficiencies and bring down costs. However, new types of renewables will also attract further interest. Researchers are now exploring technologies focused on converting waste products, such as landfill gases and wastewater treatment plant gases, to energy, and these technologies could soon offer accessibility to a wider range of renewables. Other technologies attracting attention include fast-growing, low-impact energy crops for use in combustion systems and biochemical-based hydrogen production for use in fuel cells.

Decisions on renewable energy use should determine whether the energy derived from a technology is actually renewable and whether the technology attains a life-cycle carbon-neutral status. Some renewables do generate biogenic CO_2 emissions and also produce Scope 1 and 2 anthropogenic GHG emissions, so 'renewable' is not synonymous with 'carbon neutral.'

CARBON OFFSETS AND RENEWABLE ENERGY CREDITS

In many cases, making significant investments in on-site energy production equipment or technology upgrades is not viable or cost-effective for federal leaders. Carbon offsets and renewable energy credits offer options for reducing the agency carbon emissions without large capital investments. Several government organizations have already pursued these options. For example, in 2006, EPA used renewable energy credits to become the first federal agency to displace 100 percent of its energy purchases with renewable energy. The U.S. House of Representatives has committed to being "carbon neutral" through the use of carbon offsets purchased on the CCX.

Due to the centralized production structure of electrical grids, end-users do not generally have the ability to determine the energy source that feeds their electrical outlets. Renewable energy credits allow end-users to purchase credit for certified renewable energy production regardless of the source of the electricity actually consumed. This promotes the generation of energy using renewable sources without a direct link between supplier and user. The purchase price (renewable energy credit) is passed on to the renewable energy generator.

The market and regulations for renewable energy credits have undergone rapid growth and will need to be fine-tuned and monitored in the years to come. Many states utilize renewable energy credit mechanisms to facilitate compliance with their renewable portfolio standards, which require that a certain percentage of energy produced in the state be renewable.

Similarly, carbon offsets allow consumers to purchase credits for GHG reductions made by someone else. Projects that generate offsets include reforestation, energy efficiency, and land management. These credits are the basis of the cap-and-trade programs discussed in Chapter 3, but even outside of the bounds of regulation, voluntary markets such as the CCX have sprung up to allow organizations to satisfy other stakeholder pressures and internal goals to offset GHG emissions. The prices paid by credit purchasers incentivize the development of a more robust market and technological improvements, driving down prices and facilitating the diffusion of carbon-reducing technologies throughout the market.

Weak credit processes mean the credits may not be validated. EPA and DOE can help ensure the legitimacy of credits.

Agency Sequestration Opportunities

> There is high agreement and much evidence that forest-related mitigation activities can considerably reduce emissions from sources and increase CO_2 removals by sinks at low costs, and can be designed to create synergies with adaptation and sustainable development.[2]

Up to this point, we have discussed methods for reducing GHG emissions by reducing energy consumption, especially the consumption of fossil fuel energy. Creating or increasing carbon sinks can also remove GHGs from the atmosphere. Concerted efforts in farmland, wetland, and forest management can increase storage of CO_2 and CH_4 in soils, wetlands, and plants.

Soil management methods include no-till farming, improved crop and grazing techniques, and efficient nitrogen fertilizer application to reduce emissions of N_2O, a key GHG. Wetland management involves maintaining, expanding, or creating wetland areas.[38] Forestry management practices include creating new forest lands (known as afforestation), reforesting cleared lands, and managing harvesting activities to maximize carbon storage. These techniques are often compatible, even synergistic, with other environmental conservation techniques. In fact, some of these techniques can be used to adapt vulnerable land and ecosystems to be more resistant to the impacts of climate change. For example, wetlands with CO_2-rich peaty soils can be restored in coastal areas and serve to buffer inland property against storm surges during intense weather events, which are expected to increase due to climate change.

These techniques have been used under the Kyoto Protocol and in voluntary trading schemes in the United States to create tradable carbon credits, which allow GHG-emitting activities to be offset with natural carbon sequestration efforts. Federal organizations that manage land resources should consider how those resources could be used to capture and store GHGs. The Carbon Sequestration in Terrestrial Ecosystems consortium is supported by DOE and offers guidance on opportunities for agencies.[39] Both the CCTP (see Chapter 1) and USDA also support research on terrestrial ecosystem carbon sequestration.

Economic Potential

The federal leader seeking to reduce GHGs in his or her organization is more likely concerned with the organization's budget than the macroeconomic implications of different mitigation options. However, the cost-benefit analyses for several important options make a case for action now (see *Cost-Effective Mitigation* below).

There are two broad approaches to analyzing the economic potential (tradeoff of costs and benefits) of technologies to mitigate GHG emissions. These are bottom-up and top-down approaches.

Bottom-up studies assess mitigation options, emphasizing specific technologies and regulations. For example, what is the effect of using recycled paper? The previous section describes these options. However, bottom-up studies, also described as sectoral studies, assume that the macroeconomic conditions remain static (for example, they must assume that climate change does not so affect the economy that our normal way of doing business, such as purchasing paper, changes).

Top-down studies, on the other hand, assess the economy-wide potential of mitigation options. They use a consistent framework globally and capture macroeconomic and market feedbacks. For example, a top-down approach may assume a new global cost for gasoline or electricity and ask how it affects intermodal shifts among various transportation options and the carbon emissions implications of those shifts.

As noted at the beginning of the chapter, the IPCC has concluded that both the bottom-up and top-down studies show substantial economic potential for the mitigation of global GHG emissions. This means mitigation efforts could offset or even reduce GHG emissions over the coming decades while having a positive economic impact.

However, the IPCC defines the economic potential in a very specific way as "the mitigation potential, which takes into account social costs and benefits and social discount rates, assuming that market efficiency is improved by policies and measures, and barriers are removed." This introduces a very big caveat in the opportunity since improving market efficiency is a nontrivial task in itself.

Chapter 3 discussed some of the barriers to market efficiency (for example, the political resistance to nuclear waste that reduces our ability to use nuclear energy). That feasible solutions to the mitigation challenge hinge upon these barriers is a daunting realization. As you will see next, the substantial economic potential is not enough in itself to mitigate our emissions.

Table 4-6 shows us how an office of 20 workers can save more than a ton of carbon emissions each year just from switching to 100 percent recycled paper. Now, let us assume that through regulation, carbon emissions cost \$100 per ton (Table 4-7, far right column). With that extra charge, an agency has a real economic incentive to use 100 percent recycled paper. This is the economic potential.

Table 4-7. Global Economic Mitigation Potential in 2030

<table>
<tr><th colspan="2">Carbon emissions price (U.S.$/ton$CO_2$-eq)</th><th>$0</th><th>$20</th><th>$50</th><th>$100</th></tr>
<tr><td rowspan="6">Emissions reduction</td><td rowspan="2">Economic potential ($GtCO_2$-eq/yr)</td><td rowspan="2">5-7</td><td>9–18</td><td>14–23</td><td>17–26</td></tr>
<tr><td>9–17</td><td>13–26</td><td>16–31</td></tr>
<tr><td rowspan="2">Below the high-growth scenario (%)</td><td rowspan="2">7-10</td><td>13–27</td><td>21–34</td><td>25–38</td></tr>
<tr><td>14–25</td><td>20–38</td><td>23–46</td></tr>
<tr><td rowspan="2">Below the low-growth scenario (%)</td><td rowspan="2">10-14</td><td>18–37</td><td>29–47</td><td>35–53</td></tr>
<tr><td>19–35</td><td>23–46</td><td>32–63</td></tr>
</table>

Note: The estimates are from top-down (top number in each cell) and bottom-up studies (bottom number in each cell). Data for \$0 estimates are from bottom-up studies only.

One rule—that all federal offices use 100 percent recycled paper—would save many tons of carbon emissions. The economic incentive could be there at just a few dollars per ton of emissions (that is, the extra cost of using recycled paper). Indeed, Figure 4-3 shows many options for emissions reductions that pay for themselves, while Table 4-7 suggests the mitigation potential could total 5 to 7 $GtCO_2$-eq/yr (a ton for each person on the planet) without any "price" ($0).

The next two subsections are dense, but they attempt to explain Table 4-7 and are key in grasping the challenge of meaningful mitigation.

Can Paying for Emissions Motivate Mitigation?

As in the example of the office paper, the price of carbon affects the economic potential of mitigation (in $GtCO_2$-eq/yr), as illustrated in Table 4-7. Consider just the first row of this table, economic potential. The IPCC analyses suggest reductions in global emissions by at least 9 $GtCO_2$-eq/yr (and maybe as much as 18 $GtCO_2$-eq/yr) if the price of carbon were $20 per ton, by at least 13 $GtCO_2$-eq/yr if the price were $50 per ton, and by at least 16 $GtCO_2$-eq/yr if the price were $100 per ton.

What do all these numbers mean? The world is currently emitting at least 25 $GtCO_2$-eq/yr, so dropping by 16 $GtCO_2$-eq/yr would be significant—if done right away. Of course, to do so would take time, even if the cost of emitting carbon was that much right now.

Mitigate Now or Mitigate Later?

It will take time for the effects of these prices to affect products and behaviors. During that time, emissions will continue to grow, perhaps as fast as they are now, 3 percent per year. Therefore, comparing these savings with the emissions projected for the year 2030 under two of the scenarios described in Chapter 2 (Figure 2-8) is more useful.

Scenario A1B represents a market-oriented economy with the fastest growth and a balance of energy sources, including both fossil and nonfossil fuels. This scenario ("high-growth" in Table 4-7) leads to relatively high emissions in 2030 (68 $GtCO_2$-eq/yr) though not the highest (see Figure 2-8).

The B2 scenario ("low-growth" in Table 4-7) leads to the lowest emissions of all the SRES scenarios (49 $GtCO_2$-eq/yr) though still *double* current emissions (about 7 Gt carbon or 25 $GtCO_2$-eq/yr; see Figure 1-2).

The growth of emissions will have been so great in the intervening decades that, even at a price of $100 per ton a reduction equal to our entire emission inventory today (25 $GtCO_2$-eq/yr) may only cut emissions in 2030 by 25 percent.

Our emissions today dwarf the natural system (see Chapter 1, Figure 1-3), and if we stopped emitting today, we will still have to cope with serious climate change consequences. Although the B2 scenario discussed above is the "low-growth" scenario, it represents a long-term, rapid growth of emissions. If we allow emissions to grow this much and this fast for this long into the future, the effects on the climate system are unpredictable.

Continuing status quo attitudes and policies will mean greater response costs in the future. Although political will and public demand will grow in support of mitigating actions if the impacts of climate change become more obvious, by that time the costs will increase dramatically because of the delayed response. Damage to infrastructure and ecosystems will increase with more extreme weather events, as will costs for repair and adaptation measures. In addition, the expansion of conventional technologies, such as China's construction of coal-fired power plants (at a rate of two per week), will further entrench them and impede funding their replacement.

Cost-Effective Mitigation

A study conducted by McKinsey & Company in 2008 (McKinsey study) details the costs of mitigating emissions and potential CO_2 reductions within the United States through 2030.[14] The analysis looked at three GHG abatement levels, from low-range (1.3 Gt annual reduction), mid-range (lower emissions relative to current rates), and high-range (lower emissions to meet proposed legislative goals). This study is consistent with the IPCC, in that it found that a wide array of technologies and solutions will be necessary to significantly reduce GHG emissions (see also Figure 3-4), and a large government investment will be necessary to reduce emissions to proposed stabilization levels (such as shown in Figure 3-1).

Technologies identified in the McKinsey study may reduce emissions by 3 $GtCO_2$-eq/yr in 2030, at an assumed cost of less than $50 per ton of CO_2-eq (Figure 4-3). Some important options within the jurisdiction of federal leaders are (1) improving energy efficiency of buildings and appliances, (2) encouraging higher energy efficiency in vehicles while reducing the carbon intensity of transportation fuels, and (3) expanding and enhancing carbon sinks. Strategies to implement each are presented in this chapter. Pursuit of all of them may be necessary to meet U.S. emissions targets.[14]

Figure 4-3 shows options for mitigation of GHG emissions that are available now. Those that have negative cost (meaning that they would have positive economic returns over their lifetime) are on the left. Only those on the far right of the figure cost more than $50/ton. The width of the bar provides a measure of the amount of GHG emissions mitigated by full employment of that option and the x-axis is cumulative from left to right, so the total of all mitigation measures that fall below the cost of $50/ton CO_2-eq yields a cumulative reduction of about 3 million tons. We highlighted items in the purview of the federal leader. These technologies and solutions have great potential for overcoming persistent barriers to *market efficiency*, the issue assumption discussed a few paragraphs ago.

Here the market efficiency dilemma returns in full force because the reduction of 3 $GtCO_2$-eq/yr is pretty small compared with the "economic potential" of 13–26 at $50 per ton (Table 4-7) or compared with current emissions (25 $GtCO_2$-eq/yr), let alone emissions projected to 2030. Yet these goals identified by the McKinsey study are very ambitious for current technology and resources. The United States has not adopted these options as a national strategy, but federal leaders have the challenge to articulate and encourage their adoption—and to find ways to implement them. Table 4-8 lists examples of energy reduction and GHG mitigation technologies currently being studied by various federal agencies under the CCTP.

USGCRP SAP2.1 also analyzes the mitigation costs associated with a set of GHG emission scenarios and finds significant differences among models,[40] even for the same stabilization level. Thus, long-term scenario analysis includes uncertainties that complicate the decision-making process for federal leaders and may lead to deferment of mitigation strategies unless there are co-benefits.

Another way to justify a cost for carbon emissions is through associated benefits of reductions. The IPCC points out that integrating air pollution abatement with climate change mitigation policies will potentially offer large cost reductions, compared with treating GHG mitigation policies in isolation.

Figure 4-3. GHG Reduction Opportunities Widely Distributed—2030 Mid-Range Case

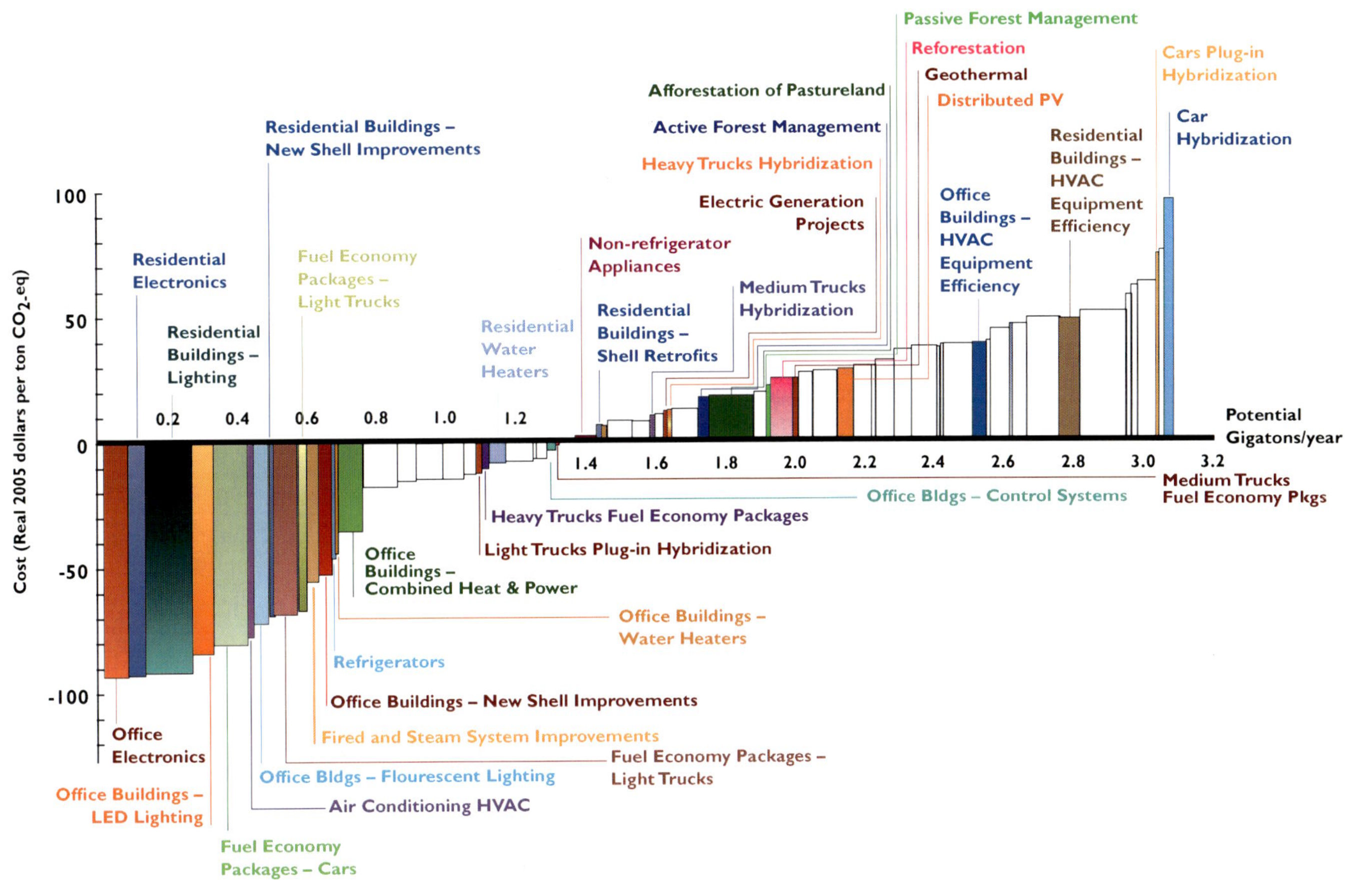

Table 4-8. Examples of Climate Change-Related Technology Research and Development Programs

USDA	Carbon Fluxes in Soils, Forests and Other Vegetation, Carbon Sequestration, Nutrient Management, Cropping Systems, Forest and Forest Products Management, Livestock, and Waste Management, Biomass Energy and Bio-based Products Development
DOC	Instrumentation, Standards, Ocean Sequestration, Decision Support Tools
DoD	Aircraft, Engines, Fuels, Trucks, Equipment, Power, Fuel Cells, Lasers, Energy Management, Basic Research
DOE	Energy Efficiency, Renewable Energy, Nuclear Fission and Fusion, Fossil Fuels and Power, Carbon Sequestration, Basic Energy Sciences, Hydrogen, Bio-Fuels, Electric Grid and Infrastructure
HHS	Environmental Sciences, Biotechnology, Genome Sequencing, Health Effects
DOI	Land, Forest, and Prairie Management, Mining, Sequestration, Geothermal Energy, Terrestrial Sequestration Technology Development
DOS	International Science and Technology Cooperation, Oceans, Environment
DOT	Aviation, Highways, Rail, Freight, Maritime, Urban Mass Transit, Transportation Systems, Efficiency and Safety
EPA	Mitigation of CO_2 and Non-CO_2 GHG Emissions through Voluntary Partnership Programs, including Energy STAR, Climate Leaders, Green Power, Combined Heat and Power, State and Local Clean Energy, Methane and High-Global Warming Potential Gases, and Transportation; GHG Emissions Inventory
NASA	Earth Observations, Measuring, Monitoring, Aviation Equipment, Operations and Infrastructure Efficiency
NSF	Geosciences, Oceans, Nanoscale Science and Engineering, Computational Sciences
USAID	International Assistance, Technology Deployment, Land Use, Human Impacts

While studies use different methodologies, there is high agreement and much evidence that in all analyzed world regions, near-term health co-benefits from reduced air pollution, as a result of actions to reduce GHG emissions, can be substantial and may offset a substantial fraction of mitigation costs.[2]

The known health effects of emissions from fossil fuel use include bronchitis, asthma, decreased lung function, and increased risk from cardiac disease. We discuss these and other health effects in Chapter 5. Countries do not report such health effects or relate them to specific sources systematically. This highlights a common challenge of environmental problems: when determining costs, managers may not incorporate the benefits of taking action.[41] Without that monetization, factoring health effects into determinations of the appropriate cost of carbon emissions is difficult.

Another way to justify a cost for carbon emissions is through associated benefits of reductions.

SUMMARY

Reducing GHG emissions poses a daunting challenge for federal leaders, but a variety of technologies are available for immediate implementation at little to no cost, or even negative cost (think recycled paper). Small organizational behavioral change can result in reductions without great cost to an agency, and numerous inexpensive technological changes can reduce emissions and energy use with real savings. The reduction in energy cost burden is a significant benefit. Leaders must choose GHG reduction measures that fit the organizational structure and budget of their organization. There is great potential for reducing emissions cost-effectively, and secondary benefits (such as health) may be notable. An educated, informed individual could implement many of the solutions described.

Although numerous technology and policy options are available for reducing energy use and GHG emissions, significant hurdles may hinder their implementation. These technologies may require large up-front capital investments compared with conventional options and may not show a return on investment until several years of energy savings accrue. Federal budgeting procedures that focus on annual expenditures rather than long-term savings do not support such capital investments. In addition, these higher capital costs may make justifying the switch from legacy systems difficult, particularly when they still have useful life remaining. What's more, because current mitigation policies are voluntary or carry few, if any, punitive consequences for failure to act, without some other impetus, agency decision makers tend to maintain existing programs.

Federal leaders' actions can have widespread benefits, but they may be difficult to quantify, due in part to the variance in benefits and costs across sectors and regions. Federal leaders can implement currently available technologies and plan for using those on the horizon. They can also play a role in long-term planning discussions regarding their agency's role in U.S. GHG reduction.

[1] National Renewable Energy Laboratory (NREL), *Energy Lab Sets Aggressive Greenhouse Gas Reduction Goal: NREL pledges to cut carbon footprint, impact on environment by 75 percent*, NREL News Release NR-2907, December 4, 2007, www.nrel.gov/news/press/2007/549.html.

[2] IPCC key finding. Throughout this book, each purple box shows a key finding from the IPCC with a high level of scientific understanding. See Appendix C for specific AR4 chapter sources for each finding. Chapter 1 gives a full discussion of our selection of key findings.

[3] Emissions to consider include the four GHGs (carbon monoxide, methane, nitrogen oxide, sulfur hexafluoride) and two groups of gases (hydrofluorocarbons and Perfluorocarbons) included in the Kyoto Protocol. See Table 1-1 for current concentrations.

[4] *The Greenhouse Gas Protocol Initiative*, 2008, www.ghgprotocol.org/.

[5] The 1605(b) had 220 reporting participants in 2005, the last year for which data are available (as of November 28, 2008); in 2008, Climate Leaders had 236 partners.

[6] The reporting protocols of the Climate Registry can be found at www.theclimateregistry.org.

[7] Regional Greenhouse Gas Initiative is a cooperative effort of 10 Northeast and Mid-Atlantic states to reduce GHG emissions and discuss the design of a regional cap-and-trade GHG reduction program. *Regional Greenhouse Gas Initiative*, www.rggi.org.

[8] ICLEI, founded in 1990 as the International Council for Local Environmental Initiatives, is the leading organization for local government action on GHG reporting. *ICLEI—Local Governments for Sustainability*, 2008, www.iclei.org/index.php?id=iclei-home&no_cache=1.

[9] The Carbon Trust, *Local Authorities —Saving Energy in Local Authority Buildings*, Sector Overview CTV028, London, UK, June 2007, www.carbontrust.co.uk/carbon/publicsector/la/.

[10] *Federal Jobs Network*, federaljobs.net/about.htm.

[11] DOE, "You Have the Power," *Federal Energy Management Program*, April 19, 2007, www1.eere.energy.gov/femp/services/yhtp/.

[12] EPA, "The ENERGY STAR Challenge," *ENERGY STAR*, www.energystar.gov/index.cfm?c=challenge.bus_challenge.

[13] EPA, *Inventory of U.S. Greenhouse Gas Emissions and Sinks: 1990–2006*, USEPA #430-R-08-005, April 2008, Table ES-2, www.epa.gov/climatechange/emissions/usinventoryreport.html.

[14] J. Creyts, A. Derkach, S. Nyquist, K. Ostrowski, and J. Stephenson, *Reducing U.S. Greenhouse Gas Emissions: How Much at What Cost?*, Executive Report of the U.S. Greenhouse Gas Abatement Mapping Initiative, McKinsey &

Company, December 2007, www.mckinsey.com/clientservice/ccsi/greenhousegas.asp.

[15] GSA, *Federal Fleet Report: Fiscal Year 2007*, January 31, 2008, www.gsa.gov/graphics/ogp/FFR2007_508.pdf.

[16] Well-to-wheels analysis includes the feedstock, fuel, and operation stages, combining "well-to-tank" emissions with "tank-to-wheels" emissions.

[17] Brinkman, Norman, Michael Wang, Trudy Weber, and Thomas Darlington, *Well-to Wheels Analysis of Advanced Fuel/Vehicle Systems—A North American Study of Energy Use, Greenhouse Gas Emissions, and Criteria Pollutant Emissions,* General Motors Corporation, Argonne National Laboratory, and Air Improvement Resource, Inc., May 2005, www.transportation.anl.gov/pdfs/TA/339.pdf.

[18] DOE, Energy Information Administration (EIA), *Annual Energy Review 2007*, DOE/EIA-0384(2007), Tables 1.2 and 2.9, June 23, 2008, www.eia.doe.gov/emeu/aer/.

[19] EPA, "ENERGY STAR For Commercial Real Estate," *ENERGY STAR*, http://www.energystar.gov/index.cfm?c=comm_real_estate.bus_comm_realestate.

[20] DOE, "EnergyPlus Energy Simulation Software," *Energy Efficiency and Renewable Energy*, November 15, 2008, www.eere.energy.gov/buildings/energyplus/.

[21] IPCC, "Summary for Policymakers" in *Climate Change 2007: Mitigation of Climate Change. Working Group III Contribution to the Fourth Assessment Report of the Intergovernmental Panel on Climate Change.* Cambridge University Press, Cambridge, United Kingdom, and New York, NY, USA, www.ipcc.ch/pdf/assessment-report/ar4/wg3/ar4-wg3-spm.pdf.

[22] DOE, "Energy-Efficient Products," *Federal Energy Management Program*, www1.eere.energy.gov/femp/procurement/eep_requirements.html.

[23] DOE and EPA, *ENERGY STAR*, www.energystar.gov/.

[24] Department of Energy (DOE), "Federal Energy Management Program," *Energy Efficiency and Renewable Energy*, www1.eere.energy.gov/femp/index.html.

[25] Department of Labor, Bureau of Labor Statistics, "Federal Government, Excluding the Postal Service," *Career Guide to Industries, 2008-09 Edition*, www.bls.gov/oco/cg/cgs041.htm.

[26] Lurita Doan [statement by Administrator, General Services Administration (GSA), before the Committee on Transportation and Infrastructure, U.S. House of Representatives, May 11, 2007], www.gsa.gov/Portal/gsa/ep/contentView.do?pf=y&noc=T&channelId=-18801&contentId=23117&programId=9641&pageTypeId=8199&contentType=GSA_BASIC; or search for "green power" on the GSA website, www.gsa.gov.

[27] For the purposes of this discussion, we consider waste management an activity conducted downstream of operational activities because the emissions that can be saved are generally realized by external entities, such as waste disposal contractors. However, certain operational changes, such as instituting an on-site food waste composting program, could result in a shifting of responsibilities from contractors to the agency.

[28] EPA, *Milestones in Garbage: A Historical Timeline of Municipal Solid Waste Management,* www.epa.gov/epaoswer/non-hw/muncpl/timeline_alt.htm.

[29] EPA, "About WasteWise," *Wastes—Partnerships—WasteWise Program*, October 7, 2008, wastewise.tms.icfi.com/wisesearch/results.asp.

[30] Mayor's Office of Operations, Office of Long-term Planning and Sustainability, *Inventory of New York City Greenhouse Gas Emissions*, Mayor of New York, April 2007, home2.nyc.gov/html/planyc2030/downloads/pdf/emissions_inventory.pdf.

[31] Federal Network for Sustainability, *USEPA Region 10—Paper Reduction Case Study*, December 2006, www.federalsustainability.org/initiatives/gfcp/EPA10papercasestudy.pdf.

[32] The Greening Federal Copier Paper Wizard is accessible at www.federalsustainability.org/initiatives/gfcp.htm, click link under section "Take Action."

[33] GAO, *Electronic Waste: Strengthening the Role of the Federal Government in Encouraging Recycling and Reuse*, GAO-06-47, November 2005, www.gao.gov/new.items/d0647.pdf.

[34] *Federal Electronics Challenge*, www.federalelectronicschallenge.net/index.htm.

[35] U.S. Fairchild Air Force Base, *Statement of Work for Base Custodial Service*, April 19, 2005, www.ofee.gov/gp/custodialsow.pdf.

[36] Office of Management and Budget, Budget Concepts Branch, *Object Class Analysis: Budget of the U.S. Government, Fiscal Year 2009,* www.whitehouse.gov/omb/budget/fy2009/pdf/objclass.pdf.

[37] The federal government can take advantage of net metering through arrangements such as power purchase agreements, which implement contracts with industry partners to manage electricity sales.

[38] Although wetlands are generally considered a carbon sink, methane production in some wetlands may provide a source of GHG emissions, making them net sources of carbon (see Figure 2-2).

[39] DOE Consortium, *Carbon Sequestration in Terrestrial Ecosystems*, csite.esd.ornl.gov/.

[40] CCSP, 2007: *Scenarios of Greenhouse Gas Emissions and Atmospheric Concentrations (Part A) and Review of Integrated Scenario Development and Application* (Part B). A Report by the U.S. CCSP and the Subcommittee on

Global Change Research [Clarke, L., J. Edmonds, J. Jacoby, H. Pitcher, J. Reilly, R. Richels, E. Parson, V. Burkett, K. Fisher-Vanden, D. Keith, L. Mearns, C. Rosenzweig, M. Webster (Authors)]. DOE, Office of Biological & Environmental Research, Washington, DC, USA, 260 pp.

[41] IPCC, "Human Health" in *Climate Change 2007: Impacts, Adaptation, and Vulnerability. Working Group II Contribution to the Fourth Assessment Report of the Intergovernmental Panel on Climate Change*, Chapter 8, Cambridge University Press, Cambridge, United Kingdom, and New York, NY, USA, www.ipcc.ch/pdf/assessment-report/ar4/wg2/ar4-wg2-chapter8.pdf.

Chapter 5
Impacts of Climate Change

Knowledge of climate change impacts increases daily, and the effort to track the literature is prohibitive for most federal leaders. In light of these limitations, a prudent first step is to rely on those appointed to that task: the IPCC is the first such resource. Because IPCC's understanding of impacts is developing even as the climate itself changes, identifying them will continue to be a challenge.

Even findings the IPCC considers very certain today may be established with greater confidence during the next assessment. For now, the prudent federal leader will rely on the impacts recognized by the IPCC with high confidence (as discussed in Chapter 1). This chapter provides an overview of these impacts and the implications for U.S. federal agencies. In Chapter 6, we describe options to adapt to climate change.

As described in Chapter 2, the earth is already irrevocably bound to experience climate change due to the increased load of GHGs humans have added to the atmosphere in the last 100 years. Depending on the policies selected in the near term (Chapter 3) and the success of mitigation efforts (Chapter 4), the amount of change and adaptation (Chapter 6) needed will be greater or less than the business-as-usual case. Climate change will affect other countries, too, and your agency will have to consider impacts and requirements abroad (Chapter 7).

MOST LIKELY IMPACTS

Preparing for proactive adaptation will require you to assess the impacts that your agency or program may face (many may not be obvious). Impacts may be observed (so the adaptation is reactive), forecast from current trends, or predicted by models using the results of climate change scenarios as described in Chapter 2. A good starting point for understanding impacts your agency will encounter is the IPCC assessment (AR4, Working Group II), in particular the impacts reported with high likelihood (Table 5-1).

> *The impacts in Table 5-1 rest on much evidence and high agreement between multiple studies. They warrant attention now.*

Table 5-1. IPCC Consensus on Future Climate Change Impacts with High Likelihood

Phenomena and direction of trend	Likelihood[a]	Agriculture, forestry, and ecosystems [4.4, 5.4]	Water resources [3.4]	Human health [8.2, 8.4]	Industry, settlement, and society [7.4]
Over most land areas, warmer and fewer cold days and nights, warmer and more frequent hot days and nights	Virtually certain	Increased yields in colder environments; decreased yields in warmer environments; increased insect outbreaks	Effects on water resources relying on snow melt; effects on some water supplies	Reduced human mortality from decreased cold exposure	Reduced energy demand for heating; increased demand for cooling; declining air quality in cities; reduced disruption to transport due to snow, ice; effects on winter tourism
Warm spells/heat waves. Frequency increases over most land areas	Very likely	Reduced yields in warmer regions due to heat stress; increased danger of wildfire	Increased water demand; water quality problems, e.g., algal blooms	Increased risk of heat-related mortality, especially for the elderly, chronically sick, very young, and socially isolated	Reduction in quality of life for people in warm areas without appropriate housing; impacts on the elderly, very young, and poor
Heavy precipitation events. Frequency increases over most areas	Very likely	Damage to crops; soil erosion, inability to cultivate land due to waterlogging of soils	Adverse effects on quality of surface and groundwater; contamination of water supply; water scarcity may be relieved	Increased risk of deaths, injuries, and infectious, respiratory, and skin diseases	Disruption of settlements, commerce, transport, and societies due to flooding; pressures on urban and rural infrastructures; loss of property

Note: Numbers in brackets refer to chapters of the AR4 WG2 report where these impacts are discussed, see Appendix B.

[a] The likelihood of future trends is based on projections for the 21st century using SRES scenarios.

The IPCC has reported other impacts that could occur, but has assigned them a greater uncertainty (see Chapter 1, where we describe these uncertainty designations along with our emphasis on the more certain findings in this book). Some of those other impacts could have a high cost (human, social, economic, etc.) if they occur, so you might consider them later as your effort to understand and respond to climate change grows in scope and capability. However, the impacts in Table 5-1 rest on much evidence and high agreement among multiple studies. They warrant attention now.

Federal agency leaders will be expected to address the changes described in Table 5-1. Below, we detail three big impacts to expect (warming days, heat spells, and heavy precipitation), but they are not the only impacts the United States will experience. Moreover, we are not certain they are the most severe impacts we will suffer, but the consensus is clear on them. Federal leaders should understand the basis of these projections and consider how they will affect the operation of their agencies.

We will discuss three specific regions of the United States that will experience these changes:

1. Coasts
2. The Western United States
3. Alaska.

Evidence suggests that people and ecosystems in these three regions are particularly vulnerable. Federal leaders with responsibilities, facilities, or operations should understand the very likely impacts in these areas and consider planning now to prepare for them. These regions are not the only ones that will experience impacts of climate change, but they are examples that are relatively well known.

Table 5-1 comes from the IPCC AR4. Other bodies also report impacts that you should consider (see Chapter 1). Most directly relevant among these is the CCSP. Several SAPs from that group deal directly with impacts in the United States, and we mention several in this chapter. Because those SAPs do not consistently report the level of confidence in findings, selecting key recommendations for immediate attention is more challenging.

THREE BIG IMPACTS

Warmer air can hold more moisture, and this physical fact creates an odd combination of effects as temperatures rise. Air able to hold more moisture will draw moisture from the ground and dry the soil. Combined with warmer temperatures, this creates more intense and longer-lasting droughts. Cooling of the moisture-laden air produces intense precipitation and damaging storms, but only when such cooling occurs (which will be rarer in a warmer climate).

It is very likely that hot extremes, heat waves, and heavy precipitation events will continue to become more frequent.[1]

Chapter 2 presents future scenarios and the modeling basis of estimated changes in temperature and precipitation. Even with small increases in mean temperature, these studies project more intense, more frequent, and longer-lasting heat waves. The daily temperature range will decrease, so nighttime temperatures will be higher with less relief from the daytime heat. The growing season will lengthen due to fewer frost days, and we can expect fewer cold air outbreaks in the eastern United States.[2]

Let's look in more detail at the three very likely impacts reported in Table 5-1.

Changing Temperature Extremes

Warming temperatures may require changes in materials, maintenance, and operations.

For instance, the combined effects of an increase in mean and extreme high temperatures in the Gulf Coast region are likely to affect the construction, maintenance, and operations of transportation infrastructure and vehicles.

Some types of infrastructure deteriorate more quickly at temperatures above 32°C (90°F). As the number of very hot days increases, different materials may be required. Further, restrictions on work crews may lengthen construction times. Rail lines may be affected by more frequent rail buckling due to an increase in daily high temperatures. Ports, maintenance facilities, and terminals are expected to require increased refrigeration and cooling.[2]

As noted in Table 5-1, the IPCC forecasts, "Over most land areas, warmer and fewer cold days and nights, warmer and more frequent hot days and nights."

This finding is confusing; why is the word warming used twice? Let's parse the sentence and consider the parts. Consider the first part, *warmer and fewer cold days and nights.* This means the cold days won't be as cold during the day. Moreover, the cold nights won't be as cold. As cold as what? As cold as the current pattern, our historical record. And it means that we won't have as many occurrences of cold days. For example, if you average 5 days at–20˚C (–5˚F) per year at your installation now, you may average 4 such days in the future.

Now let's parse the second part, *warmer and more frequent hot days and nights.* The same reasoning applies, but now instead of dealing with the cold days of winter, we are speaking about the hot days of summer. Hot days will be hotter. Hot nights will be hotter. In addition, there will be more hot days and more hot nights.

Figure 5-1 shows the historical observations and simulations of very warm nights in North America over the past century along with projections for the future under scenarios A2, A1B, and B1 (see Chapter 2). On a "very warm night," the low temperature is in the top 10 percent of warm nights.

Figure 5-1. Increase in Percentage of Very Warm Nights (top 10 percent) in North America

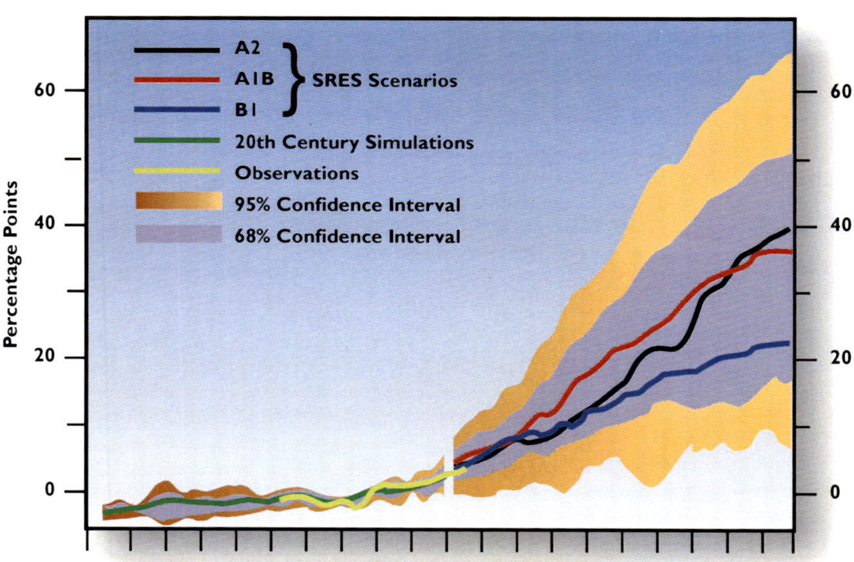

In general, four types of impacts will arise from the changes in cold days and hot days. First, fewer cold days and the fact the coldest days will tend to be less cold mean fewer human deaths from cold exposure. This will indirectly affect many agencies (for example, in productivity) and directly affect some with responsibilities for human health and welfare.

Second, warmer weather in winter can lead to reduced snowpack in mountains. In addition, warmer weather in summer can mean more water demand and less supply. This can affect agriculture and food supplies. Later in this chapter, we describe the changes like this that are now observed in the western United States and how future climate change will stress water supplies there.

The third very likely impact is less demand for heating in winter and more demand for cooling in summer. The higher temperatures also will combine with pollution to reduce air quality in cities. Moreover, less snow in winter means less impact on transportation (reduced road and track maintenance, fewer snow days). Less snow will particularly affect winter tourism (easier access to tourist areas, but less snow for recreation).

Finally, agricultural yields will be affected. In environments that are now cold, such as the northern Midwest, more warm days and fewer cold days may lead to longer growing seasons and greater agricultural productivity. Conversely, in very warm areas, the decrease in cold days and increase in warm days may lead to increased crop stress and decreased agricultural productivity. Overall, because crops vary with climate region, the variety and abundance of various food types may shift.

More Heat Waves

The IPCC also forecasts, "Warm spells/heat waves. Frequency increases over most land areas."

As temperatures rise, extremes of dryness will increase with longer dry periods and greater risk of drought. The general global pattern will be increased summer dryness and winter wetness (see Figure 7-3).

> North America—Cities that currently experience heat waves are expected to be further challenged by an increased number, intensity, and duration of heat waves during the course of the century, with potential for adverse health impacts.[1]

Further exploring the connection between increased temperatures and resulting heat waves and health impacts, the IPCC finds that North

America is particularly susceptible to damaging events. Combining these events with threats to the food and water supplies as introduced above, significant disruption in urban areas is likely to be an increasing challenge in the 21st century.

In addition to the impacts of heat waves on populations and the increased stress on populations from compromised water supplies, temperature and precipitation changes can alter the distribution of disease vectors and increase the exposure of populations to disease. For example, mosquito species that carry the West Nile virus could become prevalent in areas not currently at risk for the disease; alternatively, their hatching grounds could dry up in other areas, decreasing vector-borne disease there. In addition, stressed populations may be more susceptible to new diseases.

Temperature and precipitation changes can alter the distribution of disease vectors and increase the exposure of some populations to disease.

On the other hand, warmer winters and reduced incursions of cold air masses can be expected to reduce winter health problems. Mitigation efforts also have the potential to reduce air pollution. Chapter 4 discusses the health co-benefits of reduced air pollution. Thus, IPCC anticipates both positive and negative health impacts. These health impacts will change as climate change continues and as stress on various systems accumulates.

The balance of positive and negative health impacts will vary from one location to another, and will alter over time as temperatures continue to rise. Critically important will be factors that directly shape the health of populations such as education, health care, public health initiatives, and infrastructure and economic development.[1]

Extended periods of high temperatures also stress crops and native species and reduce yields in warmer regions due to heat stress. High temperatures directly affect plants through impacts on protein structures, and water stress reduces metabolic rates. We expect an increased danger of wildfire in natural systems as die-offs from heat and water stress lead to more tinder to start and maintain fires.

During prolonged drought, water demand of human and natural systems rises and evaporation increases. Reduced water flow in rivers leads to water quality problems, such as algal blooms and more concentrated pollution.

Increased risk of heat-related mortality, especially for the elderly, chronically sick, very young, and socially isolated is very likely, and prudent planning should consider these vulnerable populations. For example, in the European heat wave of 2003, around 60 percent of the heat wave deaths in France occurred in persons age 75 and over.

Reduction in the quality of life for people in warm areas without appropriate housing and impacts on the elderly, very young, and poor are also projected. Thus, federal agencies with responsibilities for housing and health will face new expectations.

More Big Storms

The IPCC forecasts this third impact: "Heavy precipitation events. Frequency increases over most areas."

Warmer nights and higher overall temperatures will lead to more frequent heavy rainfall. Figure 5-2 shows how heavy rainfall (expressed as a percentage anomaly from the average for the period of record) has increased over time.[4]

Figure 5-2. Periods of Heavy Rainfall Lasting at Least 90 Days in the United States, 1895–2005

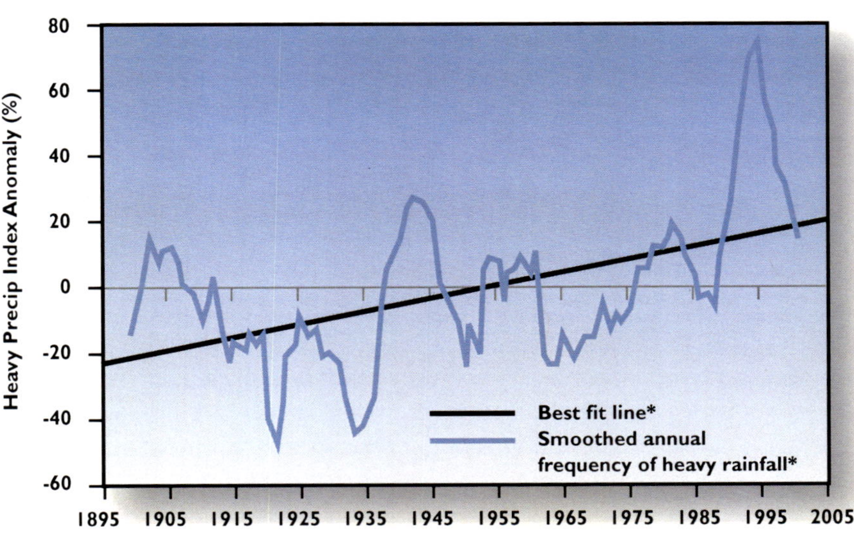

Where precipitation increases, wet extremes will be more intense and the chance of intense precipitation and flooding will increase (because warm air can hold more moisture and release more moisture). Figure 5-3

> *Wet extremes will be more intense and the chance of intense precipitation and flooding will increase.*

shows the historical observations of heavy daily precipitation in North America over the past century along with projections for the future under scenarios A2, A1B, and B1 (see Chapter 2). "Heavy daily precipitation" indicates days in the top 5 percent of precipitation. As can be seen in Figure 5-3, all three scenarios project very large increases in heavy precipitation in North America.

Figure 5-3. Increase in Percentage of Heavy Daily Precipitation in North America

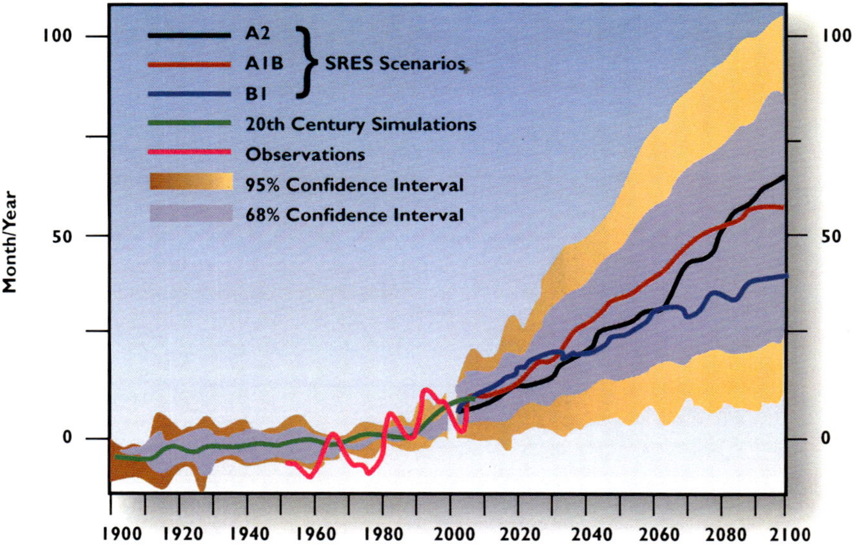

Precipitation will not increase uniformly across the continent. More northern areas, especially more mountainous ones, will tend to see increases, but the Southwest will experience less precipitation. Figure 5-4 shows the projected increases in very heavy rainfall over North America under two models. These projections are for scenarios representing CO_2 concentrations in the atmosphere that are twice current values. Heavy rainfall in these scenarios is that which exceeds 99.7 percent of the simulated daily rainfall in the control simulation (CO_2 at current values). Model 1 is the Community GCM (CGCM2) of the National Center for Atmospheric Research (Boulder, CO) and Model 2 is the HadCM3 of the Hadley Center in England. The projected increases seem to extend the observed pattern of increases in very heavy rainfall events (also shown in Figure 5-4).

Figure 5-4. Projected Increases in Very Heavy Rainfall Events in North America

Projected Increases in Very Heavy Rainfall Events (Heaviest 0.3%)

Climate Model 1

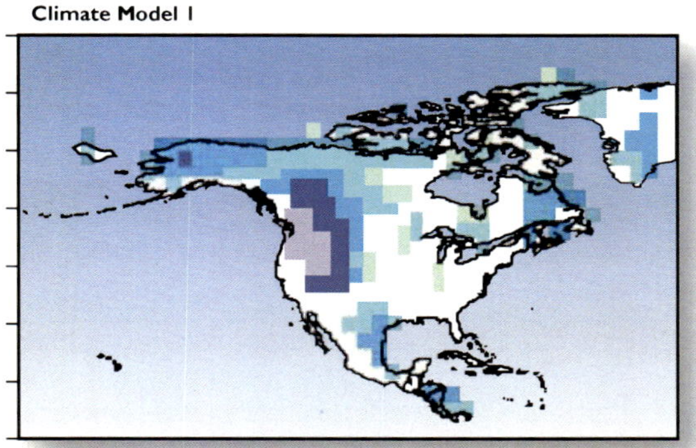

Climate Model 2

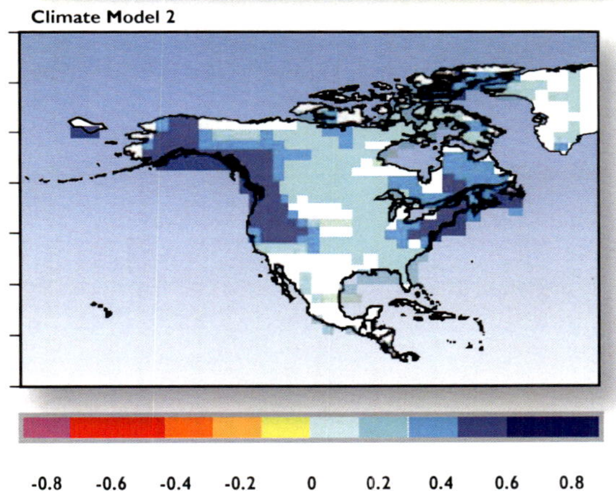

Observed Increases in Very Heavy Rainfall Events

Impacts of Climate Change

The IPCC identifies four classes of impacts very likely to arise in response to the projected increased heavy precipitation:

- Damage to crops, soil erosion, and the inability to cultivate land due to waterlogging of soils
- Adverse effects on the quality of water and contamination of the water supply, but relief from water scarcity
- Increased risk of deaths, injuries, and infectious, respiratory, and skin diseases
- Disruption of settlements, commerce, transport, and societies due to flooding, pressures on urban and rural infrastructures, loss of property.

The potential of extreme rainfall events and more frequent and prolonged flooding may disrupt traffic management, increase highway incidents, and impact airline schedules, putting additional strain on a heavily used and increasingly congested system. Further, prolonged flooding—inundation in excess of 1 week—can damage pavement substructure.[3]

One of these projected changes is positive: water scarcity may be relieved. The others are negative and may be profoundly so. Moreover, many of these changes are cumulative and mutually supporting. For example, crop damage, soil erosion, and waterlogging of soils all act to decrease food supplies. Moreover, food scarcity interacts with health problems to further weaken populations and reduce the ability to respond to other stresses such as the projected disruptions of settlements, transport, and commerce. In addition, those disruptions reduce a society's ability to respond to food scarcity and health stresses.

The IPCC considers all of these projections very likely and some virtually certain. This range of potentially serious effects can impose significant costs on the country and the operations of your agency.

Cost of Climate Change

All of these changes have the potential to significantly disrupt or burden the operations of the federal government. Higher temperatures in summer and winter affect heating loads, ecosystems, and agriculture. Particularly hard hit by more intense heat waves will be our food supply, vulnerable sick and elderly populations, federal lands, water supplies, and transportation. More intense storms produce significant damage to infrastructure and disruption of transportation and communication. Table 5-2 provides examples of effects on federal agencies.

Table 5-2. Sample Impacts on Specific Agency Operations

Phenomena	Agencies impacted (examples)
Effects on agriculture include increased crop yields in cooler areas and decreased crop yields in warmer areas, implying a restructuring of the agriculture industry in the United States. Intense storms can produce crop damage. Longer dry spells and more intense droughts can increase crop loss and the frequency and intensity of wildfires. More intense precipitation can also erode or waterlog soils, both of which reduce soil fertility.	USDA (farm disruptions) USAID (food supply changes) USFS (wildfires) DoD (help in emergencies) DHS (increased emergencies) DOI (federal lands affected) DOT (transportation disruptions)
In the Western United States, water will be less available and supplies less reliable. Demand for water will increase throughout the country during heat spells and droughts, and water quality will decline with algal blooms more frequent.	USDA (irrigation disruptions) DOE, DOI (dams and reservoirs, federal lands), EPA (water quality, species) All agencies with operations in the West
Increased risk of heat-related mortality, especially for the elderly, chronically sick, very young, and socially isolated. Increased risk of deaths, injuries and infectious, respiratory, and skin diseases.	HHS (elderly and infirm affected, disease incidence changes, adapting buildings) DHS, DoD (facilities and operations)
Lower heating costs and greater cooling costs. Air quality may decline. There will be fewer interruptions in transportation due to snow and ice storms. Winter tourism affected (for example, easier transportation and the skiing industry may suffer).	All agencies' energy costs (DOE) DOI (federal lands effected) DOT (weather disruptions on movement) HHS, EPA (pollution events)
Flooding and intense storms can disrupt towns and transportation systems and result in significant loss of life and property.	DoD (help in emergencies) DHS (increased emergencies) DOT (roads and other transportation infrastructure)
Sea level rise will expose some coastal areas to direct inundation and disrupt coastal ecosystems. Populations and infrastructure in coastal regions will be subject to storm surge with greater frequency and intensity.	DOC (shipping and ports, weather services), Housing and Urban Development (HUD) (buildings), DHS DOI, natural resource agencies EPA (endangered species) NASA (Florida launch facilities)

Current models of the costs of climate change rely on average effects and average costs country by country or sector by sector. Thus, the aggregation of costs is very coarse. More accuracy is possible through more detailed studies, considering smaller regions or populations within countries or sectors.

Models that deal at idealized individual levels may provide the greatest detail and accuracy, but they are not available now. Thus, current estimates of damage costs have weaknesses, and the coarse level of aggregation means that models may not yet have identified some economic sectors and population groups where large impacts will occur (see the IPCC finding shown below).

> It is very likely that globally aggregated figures underestimate the damage costs because they cannot include many non-quantifiable impacts. It is virtually certain that aggregate estimates of costs mask significant differences in impacts across sectors, regions, countries, and populations.[1]

Calculating future costs implies there is some specific date that is particularly important or beyond which costs can be ignored. Current understanding implies three related concerns that work against such a view. First, the climate system is dynamic and still adjusting to the effect of emissions in the past that have led to current-day concentrations of GHGs, so we are committed to certain climate changes even if emissions did not increase further. Those changes can continue for hundreds of years at least. Second, the impacts of climate change will be cumulative and accelerating as slow-reacting parts of the earth system continue to adjust to imposed GHG emissions. Third, thresholds within the earth system, once passed, may produce unanticipated shocks that ripple through the system. For example, loss of summer sea ice in the Arctic changes the amount of sunlight absorbed, further increasing water temperature, reducing winter ice pack in a positive feedback loop, and causing nonlinear behavior in the system. Thus, we have great difficulty estimating future costs because we are equally challenged to predict the future behavior of the climate system.

> In some locations and amongst some groups of people with high exposure, high sensitivity, and/or low adaptive capacity, net costs [of climate change] will be significantly larger than the global average.[1]

The IPCC estimates the global damage from climate change as the social cost of carbon (SCC). This value represents future net benefits discounted to the present (see the accompanying text box for a discussion of *discounting*).

Uncertainties in future benefits (or costs) and differences in discount assumptions lead to variations in SCC estimates (over 100 estimates) from $10 to $350 per ton of carbon, with an average value of $43 per ton of carbon.[5] Such estimates provide a basis for anticipating the value of, allocation of, and resources for adaptation. The wide divergence in these estimates ($10 to $350) implies great uncertainty and complicates the effort at cost-benefit analysis.[6]

> *Why are discount rates needed? Because a dollar received today is considered more valuable than one received in the future. Discount rates are used to compress a stream of future benefits and costs into a single present value amount. Thus, present value is the value today of a stream of payments, receipts, or costs occurring over time, as discounted through the use of an interest rate. Present value calculations of benefits and costs are then compared to determine benefit-cost ratios. Mathematically, the present value of a future benefit or cost is computed on the basis of the following equation:*
>
> $$PV = FV / (1+i)^n,$$
>
> *where PV is the present value, FV the future value, i the interest rate, and n the number of years to consider in the future.*

The current level of understanding does not allow us to predict the true future cost of climate change. Thus, efforts to adapt well in advance of the change (*proactive adaptation*, see Chapter 6) that are defensible on the basis of current cost-benefit analysis discounted to present value are justified because true costs will increase over time.

Flood and crop insurance programs must charge premiums on the basis of actuarial estimates of losses projected over long periods. Changing climate introduces special risks for such insurance projections, which are based on historical records that will not be valid in the future. GAO found that the National Flood Insurance Program and the Federal Crop Insurance Program must do more to respond to this challenge.[7]

> Impacts of climate change will vary regionally. Aggregated and discounted to the present, they are very likely to impose net annual costs, which will increase over time as global temperatures increase.[1]

The level of temporal and spatial detail in climate change scenarios now available is too coarse for many impact assessments, and the resulting estimates of costs are similarly coarse. Much greater impacts will occur in specific subscale locations, and attention to this heterogeneity using local information is justified. As described, three specific regions of the United States where significant impact is very likely are coasts, mountains in the West, and nearly all of Alaska. We discuss these three regions in detail below.

COASTS

Coastlines are particularly likely to experience the worst of climate change.

Perhaps more than any other geographic area, coastlines are particularly likely to experience the worst of climate change and are susceptible to the impacts. Over half the population of the United States now lives in the 673 counties that border a coast, and these areas are among the fastest growing (Table 5-3).

Growth is especially concentrated in coastal cities such as San Diego and Los Angeles, and 23 of the 25 most densely populated counties are coastal.[8] Density has nearly doubled in the past 4 decades and will double again in many areas during the next 2. Development along coasts tends to occur in sensitive areas, increasing the resources exposed to damage.

Table 5-3. Population Projections for United States and Its Coastal Areas

Year	U.S. Population	Coastal population
2010	299.8	158.0
2020	324.9	171.2
2025	337.8	178.0
2030	351.0	185.0
2040	377.4	198.9
2050	403.6	212.7

Many coastal cities are at low elevations and already experience damage from storm-driven tidal surges. Some coastal areas are subsiding both due to natural phenomena and compaction from increased loads of structures and dewatering as groundwater is withdrawn. This subsidence increases the risk to populations and structures. The Mississippi Delta, an example of such a system, is highly vulnerable to sea-level change and has the potential for massive population displacement.

> Coasts are projected to be exposed to increasing risks, including coastal erosion, due to climate change and sea level rise. There is very high confidence that the effect will be exacerbated by increasing human-induced pressures on coastal areas.[1]

Conversely, some coastal areas are uplifting (rebounding) from the continued effect of lightened load in formerly glaciated areas. Most of the coast of New England and Alaska as well as many areas around the Great Lakes now experience uplift. These areas are less susceptible to damage from rising sea level. The details of these changes are well known and available for planners.

Climate change threatens coastal areas from increased sea level as glaciers and icecaps melt and release water to the ocean and from rising ocean temperatures that expand ocean water. If tropical storm intensities increase (as is projected) and wind patterns change, storm surges will be more intense and damaging. This effect can become cumulative as increased storm damage leads to the erosion of protective coastal dunes and the destruction of protective structures, making coastal facilities more vulnerable to the next storm.

Sea surface temperature will increase as global average temperatures increase. This will enhance evaporation and increase moisture transport. Warmer sea surface temperatures correlate with the numbers of storms (Figure 5-5). Warmer oceans moderate land temperatures in maritime settings.

Ocean acidity is increasing as oceans absorb more CO_2 from the atmosphere. This affects the ability of shellfish to form shells and coral to form reefs. Each of these factors affects ecosystems and the infrastructure in coastal settings. Many federal agencies operate in these settings and have responsibilities that extend to these areas. In the following sections, we detail changes expected in these areas, and this inventory forms the setting in which we discuss adaptation measures in Chapter 6.

Figure 5-5. Average Number of Hurricanes/Tropical Storms and Sea Surface Temperatures for North Atlantic over Past Century

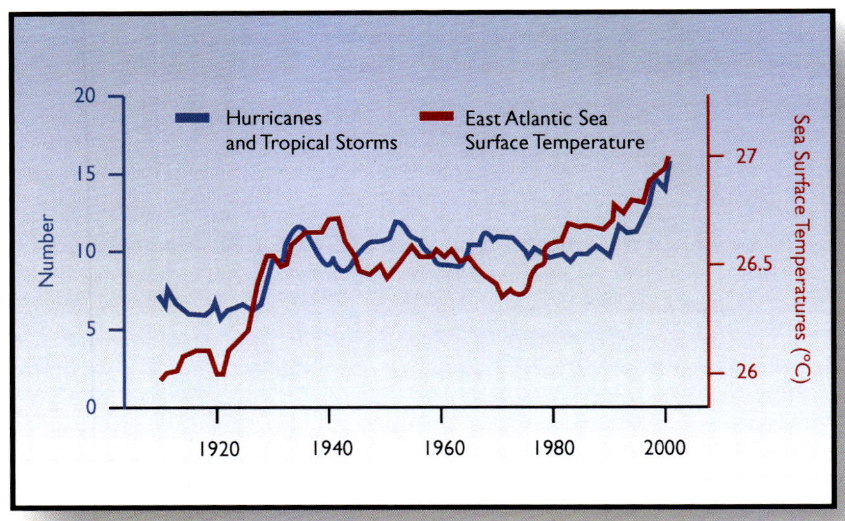

Coastal Ecosystems

Critical coastal wetlands extend the length of the Atlantic and Gulf coasts, with sporadic occurrences on the West Coast and Alaskca. These wetlands provide important nutrient sources and harbor wildlife intensively harvested for human needs. Salt marshes can moderate tidal surges and also filter pollutants transported by rivers.

Mangroves found along subtropical shorelines such as in southern Florida are especially effective at protecting shorelines from wave action, but the 240,000 square kilometers of mangrove swamps in the world represent only half of their predevelopment area. Because these wetlands are tied to the shoreline and fixed by sea level, they are particularly vulnerable to changes in sea level. If sea level rises faster than the wetlands can respond, they can be devastated by sediment accumulation, especially during storm events.

Corals are vulnerable to thermal stress and have low adaptive capacity. Increases in sea surface temperature of about 1 to 3°C are projected to result in more frequent coral bleaching events and widespread mortality, unless there is thermal adaptation or acclimatisation by corals.[1]

Corals represent a special coastal challenge in Florida and Hawaii, where they are economically important. Of the 600,000 square kilometers of coral reefs throughout the world, only 30 percent are in stable condition. Increased sediment loads from rivers, dredging, and construction are damaging the vast majority of coral reef communities. Corals contain symbiotic algae critical to the life of the coral. The algae cannot function and survive above a critical temperature specific to that group of algae. Thus, when sea surface temperatures rise above that critical temperature, the algae die and the coral loses its energy source. This thermal stress impairs growth and reproduction in the coral.

Thermal stress impairs growth and reproduction in the coral.

Another profound and demonstrable coastal (and global ocean) change independent of direct climate change is the ongoing increased acidification of the surface oceans due to uptake of CO_2 from the atmosphere (just as the soda we drink is acidic from the dissolved "carbonation" from CO_2). The oceans serve as one of the largest sinks or reservoirs of CO_2 emissions (see Chapter 2). This acidification can chemically attack shelled organisms living in the ocean water. Under model projections of the future, calcium carbonate saturation levels are very unlikely to provide adequate support for coral reefs in any U.S. water.[9]

Because corals are critical parts of the marine ecosystem and essential barriers to coastal erosion in many tropical areas (including the Florida Keys, the state of Hawaii, and several U.S. territories), impacts on corals threaten some U.S. coastal ecosystems. Some corals may be able to acclimate by expelling one algae group and entraining a different group that can survive the higher temperatures.

There is very high confidence that coastal wetlands including salt marshes and mangroves are projected to be negatively affected by sea level rise especially where they are constrained on their landward side, or starved of sediment.[1]

Wetlands respond to sea level rise by dynamic adjustments, rising by accumulating sediment and retreating landward due to the low slopes on many shorelines. The retreat may be multiples of the amount of sea level rise. If the landward side is constrained naturally (such as by cliffs) or artificially (by engineered structures), the wetland may be inundated and destroyed. If sediment is not available to accumulate in the wetland, it cannot rise with sea level and will be destroyed. Conversely, the increased deposition of sediment in the wetland may starve shoreline features down-coast, leading to erosion there. Thus, critical wetlands are particularly vulnerable to sea level rise.

Certain federal agencies may operate in wetlands or have responsibility for wetland protection or restoration. These agencies can expect significant challenges in fulfilling mandates or adjusting operations in response to coastal wetland damage.

Coastal Infrastructure

Of the three major changes in coastal systems (sea level rise, increased temperature, and ocean acidity), the greatest impacts on coastal infrastructure will come from sea level rise. Direct inundation of facilities and undermining of foundations will damage infrastructure at the coast. More intense storms will compound the damage by generating greater storm surge, broader damage zones, and prolonged inundation.

The greatest impacts on coastal infrastructure will come from sea level rise.

Global sea level has already risen by 10–20 cm in the past century, and models suggest this rise is very likely to accelerate. The best estimate is that the sea level will rise by an additional 48 cm by 2100, with an uncertainty range of 13–95 cm.[9]

Sea level rise is expected to exacerbate inundation, storm surge, erosion, and other coastal hazards, thus threatening vital infrastructure, settlements, and facilities that support the livelihood of island communities.[1]

The EPA estimates that shoreline retreat from a 1-meter rise in sea level would cost the United States $270 to $475 billion (Table 5-4). Those estimates ignore the impacts not readily quantified or foreseen. Table 5-4 shows the amount of land loss (area and percentage) under three scenarios of sea level rise (50, 100, and 200 cm). Also shown are costs associated with these losses. The total cost line includes the total value of land lost and the total cost of coastal defense. As the amount spent on shoreline defense rises, the amount of loss declines, so the ranges given in the total cost line reflect that tradeoff and are not simply sums of the two other amounts.

Another cost of climate change in the coastal areas comes from fisheries. In the United States, the total economic contribution of recreational and commercial fishing is approximately $40 billion per year. Climate change is very likely to substantially alter the distribution and abundance of major fish stocks.[9]

Of the $19 trillion value of all insured residential and commercial property in the states exposed to North Atlantic hurricanes, $7.2 trillion (41 percent) is located in coastal counties. This economic value includes 79 percent of the property in Florida, 63 percent in New York, and 61 percent in Connecticut.[10] High property values in these areas also reflect the growing proportion of seasonal homes.

The major direct impacts of human activities exceed any observed influences of climate change, including sea level rise. Examples of human impacts include the drainage of coastal wetlands and discharge of sewage, fertilizers, and contaminants into coastal waters. Extracting sand and hydrocarbons, fishing, and invasive species already stress coastal ecosystems. Construction of seawalls, damming, channelization, foredune construction, beach nourishment, and diversions of coastal waterways change the coast and circulation patterns as well as nutrient and sediment loads. These impacts will make it harder for both humans and the natural environment to adapt to climate change.

Table 5-4 shows another example where multiple factors (increasing human density in coastal regions, sea level rise, storm risk, etc.) are important over the full planning horizon. Negative feedback factors (such as the inability to obtain insurance in at-risk coastal regions) may mitigate these risks over the planning horizon. Incorporating these different influences will be a challenge for agencies as scenarios and adaptation strategies are developed.

Western Water Resources

Water supplies in the western United States differ from those in the East because their primary source is seasonal melting of snowpack, which accumulates in winter storms in the higher mountains. Nearly all spring and summer flows derive directly from snowpack, which constitutes the major "reservoir" of the West. Water stored in snow is essential for maintaining year-round flows and the melting of snows at higher elevations as summer progresses ensures water supplies. This snowpack is extremely vulnerable to climate change and is already impacted by warming temperatures.

Table 5-4. Land and Economic Loss from Sea Level Rise

Category	Baseline	50 cm	100 cm	200 cm
If no shores are protected				
Dryland lost (km^2)	NC	1,280–2,822	1,977–3,987	3,162–5,942
Wetlands lost (%)	NC	17–43	26–66	29–76
If developed areas are protected				
Dryland lost (km^2)	1,470–4,686	849–2,355	1,583–3,551	2,470–5,211
Wetlands lost (%)	9–25	20–45	29–69	33–80
Value of lost land[a]	24–71	79–197	130–321	170–450
Wetlands	8–65	17–124	26–194	29–218
Undeveloped land	9–29	20–51	32–108	44–183
Land for dikes	0	14–50	21–73	33–112
Cost of coastal defense	6	83–186	217–462	609–977
Open coast: sand	6	23–123	41–221	89–430
Open coast: elevate structures	0	44–55	94–257	389–479
Sheltered shores: dikes	0	8–20	17–50	45–153
Total cost	30–77	194–351	409–719	872–1,333
If all shores are protected				
Wetlands lost (%)	NC	38–61	50–82	66–90

Note: km = kilometer; NC = not calculated.

[a] Estimated economic losses displayed in billion dollars, converted from 1988 to 2008 estimated dollars.

There is very high confidence that areas in which runoff is projected to decline face a reduction in the value of the services provided by water resources.[1]

Higher temperatures have a compound effect on the snowpack-snow melt system. Warmer temperatures in fall, when snowpack normally begins to accumulate, mean that precipitation falls as rain at elevations that would normally experience snow. It is longer into the fall before temperatures drop enough to support snow formation.

Rain-on-snow events in fall become more common as temperatures fluctuate around the freezing point. Rain-on-snow events are the most serious sources of devastating flooding and can be expected to increase as the climate changes. They produce snowpack melting in fall;

snowpack cannot accumulate as quickly or develop as deep, and snow and melting of snow decline. The snowpack going into winter is thinner and less extensive.

When spring melting begins, the smaller snow reserves deplete quicker. Equally serious, spring melting begins earlier (as global temperatures rise), so the flows may take place before the principal use (agriculture) requires the water. Because melting begins earlier and the snowpack is smaller, snow is depleted earlier in the growing season and flows may not be available when required for irrigation in the heat of summer.

> *Snowpack supplies essential water resources and significant tourism opportunities in all western states.*

Snowpack supplies essential water resources and significant tourism opportunities in all western states, and many western cities depend on snowmelt-fed water flows to sustain their populations. These flows are also critical for agriculture, especially irrigated lands, for ranching and for many industries. Flows in the Columbia and Colorado Rivers as well as tributaries in the Sierra Nevada of California create hydroelectric energy that is a key renewable resource.

Competing interests already manage and apportion most of these flows, and management compacts may assume certain minimum flows to satisfy all demands. In much of the western United States, water resources are already over-allocated because of interstate compacts where allocation was based on over-optimistic estimates of average river flows and because of competing uses of water within states. In addition to overuse of surface water, groundwater is mined and depleted rather than consumed in a sustainable way.

Higher temperatures allow many species to flourish, including insects that infest forest timber. Higher temperatures also correlate with increased drought, which can stress trees and make them more susceptible to pests. The combination of insect-stressed trees and dry conditions leads to increased flammable material on the forest floor, which is ripe for intense forest fires and destruction of large tracts of forest.

Parts of the West, especially the Southwest, have been suffering through drought for a number of years, and this drought has stressed the ecosystems and resulted in significant die-off of the dominant species of trees. The bark beetle is severely damaging stands of junipers, which dominate many areas in this region. A similar infestation has affected several species of pine (lodgepole, ponderosa, and white-barked) throughout the Rocky Mountains. Dead trees form the perfect setting for severe forest fires, which have grown in frequency and intensity. As the climate changes, conditions favoring a continuation of this pattern are expected. Recent fires in Los Angeles illustrate a situation that may become more commonplace.

> North America—Disturbances from pests, diseases, and fire are projected with very high confidence to have increasing impacts on forests, with an extended period of high fire risk and large increases in area burned.[1]

Climate change will exacerbate these problems, challenging federal facilities and lands in the West. Key elements of the water cycle that can be expected to change or become more variable include

- precipitation amount;
- precipitation timing, frequency, and intensity;
- evaporation and transpiration;
- average annual runoff; and
- snowpack.

Climate change will also affect water quality. For example, more frequent or more intense forest fires resulting from droughts can lead to increased soil erosion, which degrades the water supplies reaching treatment plants.

> North America—Warming in western mountains is projected with very high confidence to cause decreased snowpack, more winter flooding, and reduced summer flows, exacerbating competition for over-allocated water resources.[1]

At the forefront of efforts to predict water flow conditions in large river basins are regional models integrated with the global climate models (see Chapter 2), which calculate climate change from emission scenarios. Because the coarse resolution of global climate models does not allow representation of the elevation variations in mountains that determine snow distribution and the snow-rain boundary, the regional climate results are used in hydrologic models to downscale the results to meaningful spatial resolutions. Results of these models are only as good as the input climate solutions, and significant difficulties remain in representing precipitation and computing runoff and stream flow. The models continue to advance and may form the basis of estimates of the many related natural and social processes responding to climate change.

ALASKA

The clearest impacts of climate change occur in polar areas, including northern parts of Alaska. All global climate models project that the surface temperature will increase the most in polar areas. They consistently indicate that this region will greatly change in response to increasing global temperatures. These changes include melting glacier ice and sea ice, increased precipitation, greater absorption of solar energy, major ecosystem changes, and enhanced release of GHGs.

The clearest impacts of climate change occur in polar areas, including northern parts of Alaska.

Increases in the amount of precipitation are very likely in high latitudes.[1]

Higher precipitation amounts and moister soil increase biological productivity in polar areas such as Alaska. Together with significantly higher temperatures, these climate changes will produce shifts in the distribution of plant species (Figure 5-6) and consequent adjustments in animal distributions. The Arctic is becoming greener, and the exchange of GHGs with the atmosphere is increasing.

Surface albedo (amount of light reflected) is decreasing in Arctic areas. The greener (and darker) surface absorbs more solar radiation, and this leads to warmer soil temperatures. As soil warms, permafrost melts. This releases gases stored in the frozen ground and is happening at a large scale and accelerating pace that could release enormous quantities of CH_4, a GHG more potent than CO_2 (Table 1-1).

Thawing permafrost in Alaska is already creating settlement and land subsidence problems for roads, rail lines, runways, and pipelines.[11]

Figure 5-6. Vegetation of Arctic: Current Conditions and Projected Changes under IS92a Scenario (2090–2100)

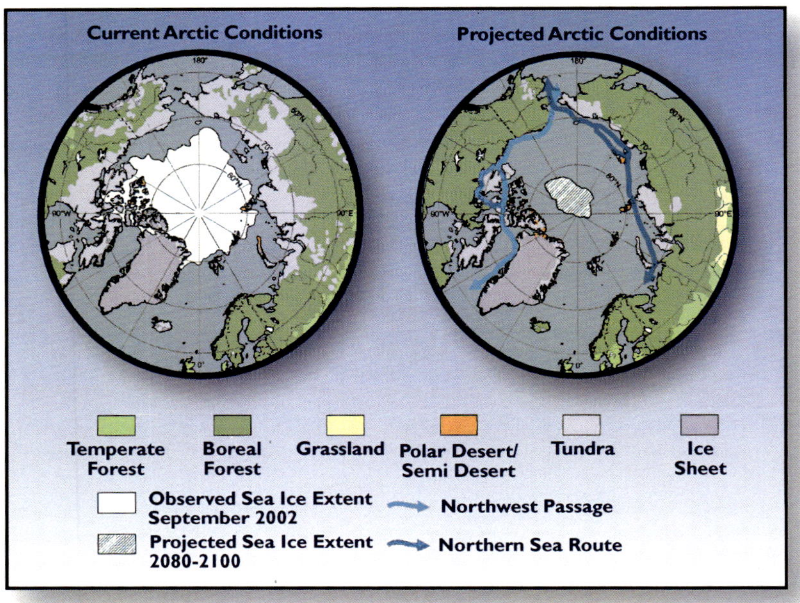

Higher precipitation amounts also generate greater river flows. Fresh water now enters the Arctic Ocean and other polar seas in greater amounts than before. This (low-salinity water) can destabilize thermohaline circulation, which is delicately dependent upon the density of salty ocean water.

Higher Arctic Ocean temperatures and greater freshwater runoff both enhance the melting of sea ice. Traditionally, sea ice extent in the Arctic Ocean reaches its lowest level in September each year. Figure 5-7 shows how September ice extent has declined over the past 3 decades. The rate of decline in this period is 10 percent per decade, or 72,000 km^2 per year. The decline in 2007 was 23 percent lower than the previous record low observed in 2005 (Figure 5-8). Ice-dependent marine wildlife (seals, polar bears, plankton, etc.) is adversely affected by the loss of sea ice.

For those managing in Arctic regions, this can mean more open seas (Figure 5-6) for transportation in summer but increased challenges for energy exploration because land transport is disrupted as the permafrost melts and open ocean conditions challenge drilling rig stability. Melting permafrost also disrupts structures and roads, which will require significant investment to rebuild and relocate.

Figure 5-7. September Arctic Ice Extent from 1979 to 2007

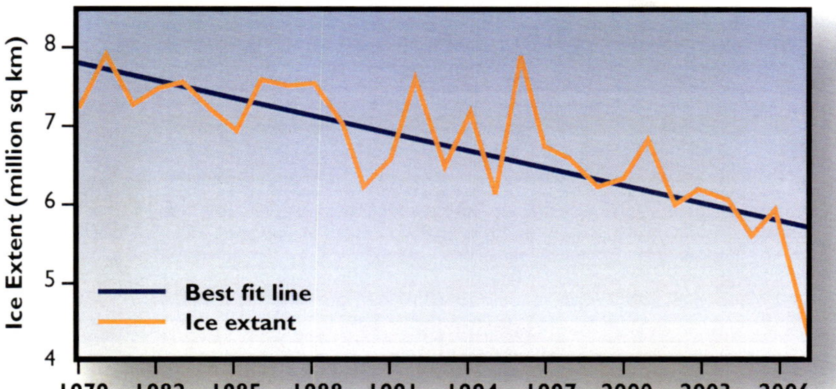

Melting of Arctic sea ice exposes shorelines to wave action and erosion. Increased wave action—due to seas that are more open and greater fetch, as well as stronger winds accompanying more intense storms—results in greater erosion of shorelines and damages coastal facilities.

Figure 5-8. Spatial Distribution of Arctic Sea Ice in September 2007

Positive impacts expected in Arctic areas include increased use of ocean transport systems and extended tourism seasons. Increased biological activity and greater agricultural productivity may benefit local populations.

Agency Perspective

All agencies will find it important to consider the potential impacts of climate change and to be proactive in preparing to deal with the effects, wherever their offices are located.

Among other functions, the Treasury needs to print money, mint coins, get them to a Federal Reserve Bank, and assist taxpayers. Homeland Security (FEMA, Coast Guard, Transportation Security Administration, Customs, Secret Service, and Border Patrol Protection) has significant public safety missions to carry out regardless of climate change impacts.

Each agency head and senior staff member may want to think of this issue in a holistic manner, not as just another emergency preparedness or business resumption planning exercise. What are the potential effects of climate change on the organization's customers and clients, and how can it still provide services in abnormal situations? What are the impacts—not just to the facility and staff, but the public as well—if appropriate proactive measures are *not* taken?

Summary

In addition to the changes in climate already observed over the past 100 years (Chapter 2), the increasing levels of GHGs in the atmosphere are reason to project greater changes in climate in the future. Three classes of change in climate are considered almost certain or very likely.

The IPCC has found that it is virtually certain that extremes of daily temperatures will get warmer. Low temperature extremes in winter will not be as low and high temperature extremes in summer will be higher. High temperature extremes produce heat stress and other health problems. High temperatures are also associated with increased water demand and decreased agricultural productivity. These heat extremes also can add to air-conditioning expense in summer. Longer and more intense heat waves, which magnify the effects of the daily heat extremes, are very likely.

Larger and more intense storms are also very likely, which increase the frequency of flooding, lead to crop damage, exacerbate health problems, and disrupt transportation systems.

Other impacts are likely but less certain than the three mentioned above. Some of them may be more severe, and we expect to better understand them in the coming years. These various impacts will add significant costs to the operation of federal agencies. These costs will stem from impacts on the internal operations of the agency, availability of staff members, and increased responsibilities for responding to the impacts of climate change, nationally and internationally.

The three geographic areas where impacts are projected to be particularly challenging include coastal counties, where over half the population of the United States now lives; the western United States, which depends on water supplies from mountain snowpack; and Alaska, which has already experienced significant impact due to warming in the past few decades.

Climate change will affect all agencies, and the process of identifying potential impacts will be complex and lengthy. We expect that impacts will vary geographically and over time and will affect many parts of the climate and human systems in many different ways. Assessment of impacts will be a continuing process. As climate changes and as our mitigation decisions define our path of change, we will know more about the changes in store.

---

[1] IPCC key finding. Throughout this book, each purple box shows a key finding from the IPCC with a high level of scientific understanding. See Appendix C for specific AR4 chapter sources for each finding. Chapter 1 gives a full discussion of our selection of key findings.

[2] With a warmer Arctic and warmer Gulf Stream, fewer cold air incursions from Canada will occur.

[3] U.S. CCSP, *Impacts of Climate Change and Variability on Transportation Systems and Infrastructure: Gulf Coast Study, Phase I*, Synthesis and Assessment Product 4.7, March 2008, pp. ES-5 and ES-6 www.climatescience.gov/Library/sap/sap4-7/final-report/sap4-7-final-exec-sum.pdf.

[4] The data reported are the excess precipitation in a 90-day period that exceeds the U.S. average 1 in 20-year event. The line shows the best-fit trend.

[5] IPCC, *Climate Change 2007: Impacts, Adaptation, and Vulnerability. Working Group II Contribution to the Fourth Assessment Report of the Intergovernmental Panel on Climate Change*, Chapter 20, Section 20.6, Cambridge University Press, Cambridge, United Kingdom, and New York, NY, USA, www.ipcc.ch/pdf/assessment-report/ar4/wg2/ar4-wg2-chapter20.pdf.

[6] NOAA Coastal Services Center, "Discounting and Time Preference," *Restoration Economics*, www.csc.noaa.gov/coastal/economics/discounting.htm.

[7] GAO, *Climate Change: Financial Risks to Federal and Private Insurers in Coming Decades Are Potentially Significant*, Report to the Committee on Homeland Security and Governmental Affairs, U.S. Senate, by John B. Stephenson et al., GAO-07-285, March 2007, www.gao.gov/new.items/d07285.pdf.

[8] Department of Commerce, NOAA, and the National Ocean Service, *Population Trends Along the United States Coast, 1980–2008*, Coastal Trends Report Series, 2005, oceanservice.noaa.gov/programs/mb/supp_cstl_population.html.

[9] USGCRP, National Assessment Synthesis Team, "Overview: Coastal Areas and Marine Resources," *Climate Change Impacts on the United States: The Potential Consequences of Climate Variability and Change,* 2000, www.usgcrp.gov/usgcrp/Library/nationalassessment/overviewcoastal.htm.

[10] IPCC, "Summary for Policymakers" in *Climate Change 2007: Impacts, Adaptation, and Vulnerability. Working Group II Contribution to the Fourth Assessment Report of the Intergovernmental Panel on Climate Change*, Chapter 20, Section 20.6, Cambridge University Press, Cambridge, United Kingdom, and New York, NY, USA, www.ipcc.ch/ipccreports/ar4-wg2.htm.

[11] *National Academy of Sciences, Transportation Research Board,* The Potential Impacts of Climate Change on U.S. Transportation, *Special Report 290, pp. 22–23.*

Chapter 6
Adaptation to Climate Change

YOUR ROLE

Regardless of the course of action that world political leaders take over the next few decades, the world's climate conditions have already been altered enough to cause long-term changes in climate (temperature, precipitation, etc.) in certain regions of the United States. Even if the world significantly reduces GHG emissions (and there is no current consensus), the lag in the climate system (see Chapter 2) means that we are faced with decades of climate change due to the GHGs already in the atmosphere. The courses of action available to the current generation of U.S. political leaders aim primarily to mitigate the extremes of those changes without causing unacceptable disruptions to the American public's quality of life.

> ***Adaptation**: Adjustment in natural or human systems in response to actual or expected climatic stimuli or their effects, which moderates harm or exploits beneficial opportunities. Various types of adaptation can be distinguished, including anticipatory, autonomous, and planned adaptation.*

What does this mean for federal leaders today? In short, it means you will have to begin to prepare your agency operations and programs to adapt to the changing conditions of the next 15, 25, or 50 years. Some agencies already look as far ahead as 50 years when planning. For example, the standard Army Corps of Engineers protocol calls for the use of 50- and 100-year periods for cost-benefit studies. For other agencies, use of such planning horizons will require new strategies and the development of new mechanisms. Adaptation, or more specifically, proactive adaptive management, is a framework for achieving agency mission objectives in the face of the unavoidable and still uncertain impacts of climate change.

Beyond preparing your agency, federal leaders have another opportunity—to lead. Leadership by federal agencies on adaptation is essential, particularly at this early stage in understanding and preparing for the impacts of climate change. National, state, and local government agencies have differing and complementary roles in climate change adaptation, but federal agencies often chart the course for the following:

- Developing, implementing, and reviewing policies and strategies, including regulations, standards, and market incentives
- Integrating climate change considerations into existing agency policies, strategic planning processes, and operations
- Building adaptive capacity, including providing tools and information, raising awareness of adaptation options, and educating key groups about adaptation
- Investing in climate change science as well as related social, ecological, and economic studies
- Managing risks from climate change to their own operations, programs, and assets, including natural ecosystems (federal lands, national forests and parks, and coastal zones) and infrastructure (highways, navigation channels, water reservoirs, and electricity grids).

A proactive adaptive management framework also recognizes that those who understand the context and likely consequences of action are best suited to manage risk. As such, federal and state government leaders need to learn the probable impacts of climate change.

Due to the complex nature and global scale of climate change, the government cannot cope with it alone. Businesses and nongovernmental organizations have an important role in adapting to climate change risk, and government leaders will have to engage these partners throughout the process.

As with the other chapters in this guide, we provide a framework for federal leaders to integrate efforts to adapt to climate changes into their long-term strategic plans for agency operations and resource stewardship. We also provide cases where government agencies are already taking action—at the federal, state, and international levels. Finally, we point you to where you can learn more.

What Is Climate Change Adaptation?

Climate change adaptation refers to any action that reduces the negative impact of climate or takes advantage of potential new opportunities. Proactive adaptation is action implemented prior to a climate-driven impact. It is most effective as a guiding principle when several options are under consideration. Reactive adaptation occurs in response to the impact of a climate event and most commonly takes place after a natural disaster. In most cases, proactive adaptation is more cost-effective and efficient. Still, our ability to forecast climate change is limited: we can expect to make errors, bringing surprises that require corrections.

This means some of your adaptation efforts will be reactive despite your best efforts. Reserves to handle such reactive adaptations must be part of your planning. Another aspect of adaptation planning is recognizing and reducing vulnerability to climate change. Although you cannot fully know the nature of climate change and its impacts, you may know enough to identify and reduce vulnerabilities; this is the second major aspect of proactive adaptation.

Table 6-1 describes six typical categories of adaptation measures. In most cases, the aim of adaptation is to prevent losses and increase the resilience of an organization or a program to conduct its mission given changing climate conditions.

Table 6-1. Climate Change Adaptation Categories

Category of adaptation	Strategy	Examples
React after impacts	Do nothing to reduce vulnerability and absorb losses	Rebuild or abandon affected structures after an extreme storm event
Enhance built environment to reduce or prevent impacts	Adapt built systems to reduce vulnerability	Engineer structures to withstand greater wind, heavier precipitation, and more frequent flooding
Spread or share the cost of reacting to impacts	Spread burden of costs of impacts across different systems or populations	Provide subsidized flood insurance or catastrophic loss coverage
Change activity to accommodate impacts or stop it altogether	Stop or change activities that are cost prohibitive under the new conditions at a given location	Prevent development on coastal land below a set elevation and rehabilitate natural vegetation
Change the location to accommodate activity	Move an activity to a location with conditions that will better accommodate activity	Relocate coastal infrastructure further inland outside of risk zone
Enhance natural capacity of system and participants	Enhance the resiliency of the natural system to improve its ability to deal with stress	Preserve or rehabilitate natural coastal systems that protect the shore

Clearly, each federal agency will have to employ some or all of these adaptation strategies. Task forces you create to evaluate appropriate strategies will identify the relevant issues and recommend responses appropriate for the types of impacts your agency will experience in the locations in which it operates.

Consider the example of health stresses from climate change discussed in Chapter 5. Several factors mentioned in Table 6-1 may be under your control as a federal leader and can affect how well your stakeholders can withstand health-related stresses. Therefore, prudent preparation (proactive adaptation) for climate change impacts will address these challenges early through education and initiatives to prepare infrastructure and review planning regularly. Responsibility does not rest solely with the healthcare sector because individual federal leaders can impact the heat stresses their employees and other stakeholders experience through the operations of their buildings and related infrastructure.

There are also imperatives to help social systems adapt to reduce vulnerability. A leader will see the value of improved social fabric, warning systems, trust, and other intangibles in preparing society to face the challenges of adaptation to climate change.

In adapting to climate change, agency leaders will be asked to identify the programs susceptible to impacts (see Chapter 5), including particularly vulnerable systems, assess the agency's ability to adapt to these changes, and determine cost-effective actions to reduce the vulnerability of the systems. This will require identification of relevant stakeholders, regular and repeated risk analysis, development of risk management strategies, and allocation of resources to selected adaptation strategies.

Currently assessed impacts of climate change may become more severe and extensive as the magnitude of climate change increases. Figure 6-1 shows impacts that are very likely to occur. They are shown over a range of increasing temperature changes, which illustrates the increased severity in impacts we will see if we follow our present course.

The magnitude of climate change and its impacts depend, in part, on mitigation plans implemented in the next few years. Therefore, a part of the agency leader's task is to understand alternative scenarios and resulting projected trajectories of climate change (see Chapter 2) and prepare impact assessments at various planning horizons, which may extend well beyond an agency's typical strategic planning horizon.

Figure 6-1. Examples of Global Impacts Projected for Changes in Climate

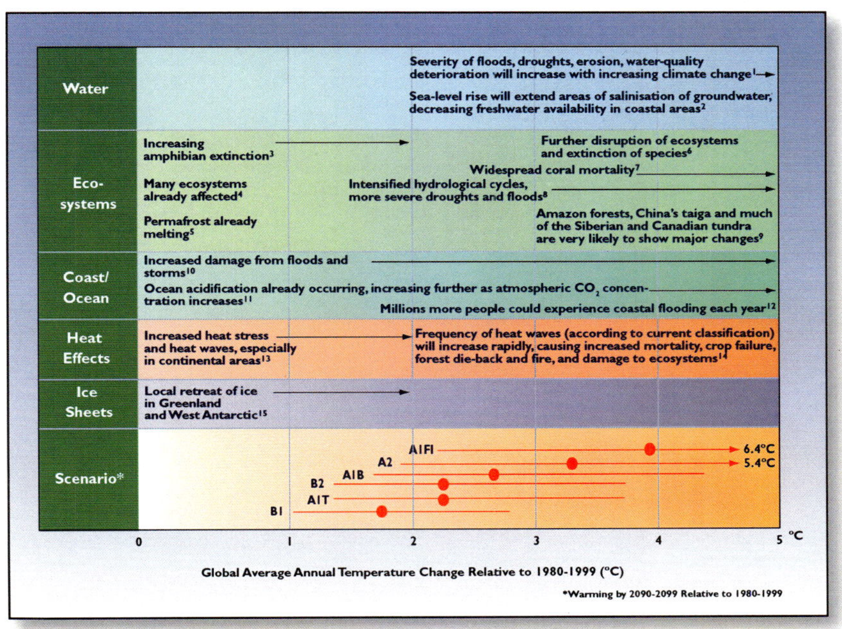

Note: This is a selection of current IPCC estimates of impacts associated with projected global average surface warming in the 21st century. Placement of findings indicate the range of temperature change to which the impacts relate. The number following each impact indicates its source; these numbers are listed at the back of the guide, in Appendix B. Confidence levels for all statements are very high, unless otherwise noted (see Appendix B).

As you become familiar with the topic and the work identifying potential impacts, you should explore the potential impacts that may be less likely to occur. Moreover, you need to recognize that some impacts unknown today may be of greater consequence than others. Some of the impacts we expect may never occur. All the uncertainties and errors in preparing for the future that have always plagued humans will bedevil us as well in the case of climate change.

To illustrate adaptation challenges and responses, we focus on two topics, coastal regions and water supply, both of which already are vulnerable and are expected to become more so in the future.

Coastal Regions Adaptation

Figure 6-2 shows the expected number of people in areas affected by coastal flooding in the next decade, mid-century, and at the end of the century, as implied by four of the SRES scenarios (see Figure 2-8). The light blue bars represent the estimates without predicted sea level rise, while the purple bars indicate those at risk when sea level rise also is included.[1] The sea level rise component represents an effect due to climate change, but most of the difference is due to changes in the number of people in coastal areas and their exposure and vulnerability to flooding as influenced by non-climatic driving forces in the scenario.

Figure 6-2. Number of People at Risk Globally from Coastal Flooding (Annually in Millions)

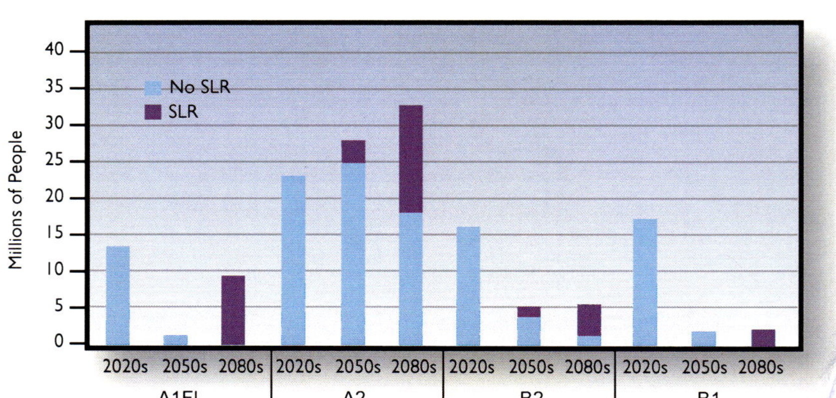

The A2 scenario projects the greatest increase in global temperature and the B1 scenario projects the least amount of global temperature increase over the next century (see Figure 2-10). If we reduce our emissions (for example, choosing B2 instead of A2), millions fewer people will be at risk from coastal flooding such as that recently experienced with Hurricane Katrina or Cyclone Nargis (see Chapter 7), but most of this difference arises from differences in vulnerability implied by the scenario, not differences in climate.

All four scenarios predict millions more people at risk from flooding due to sea level rise in the decades ahead. Three predict a drop-off between the 2020s and the 2050s. The A2 scenario predicts many more people at risk than the other scenarios because it is characterized by relatively low economic but high population growth.

The complexity the scenarios encapsulate is clear. All contemplated futures suggest that the socioeconomic factors at work will lead to increased vulnerable populations regardless of climate change and mitigation strategies selected, but climate change exacerbates the impact. This is typical of many adaptation issues. Leaders will face increased demands, amplified by climate change.

North America—There is very high confidence that coastal communities and habitats will be increasingly stressed by climate change impacts interacting with development and pollution.[2]

There is very high confidence that: population growth and the rising value of infrastructure in coastal areas increase vulnerability to climate variability and future climate change, with losses projected to increase if the intensity of tropical storms increases. Current adaptation is uneven and readiness for increased exposure is low.[2]

Considering it another way, the kinds of adaptations required for many impacts from climate change correspond to those that will be required from other sources, such as climate variability. The added effect of climate change is to increase the demand for these actions, increase the magnitude of adaptation required, or hasten the need for adaptation.

There is very high confidence that adaptation to climate change is likely to benefit from experience gained in reaction to extreme climate events, specifically by implementing proactive climate change risk management adaptation plans.[2]

Figure 6-3 shows the energy, as measured by the power dissipation index (PDI),[3] number of deaths, and inflation-adjusted economic damages from hurricanes making landfall in the continental United States. Whether recent patterns of storm damage reflect changes in climate is unclear.

Regardless of source, the responses required to protect human life and property from coastal storms and sea level rise fall broadly into three categories—retreat, accommodation, and protection:

- Retreat involves no effort to protect the land from the sea. The coastal zone is abandoned and ecosystems shift inward. Excessive economic or environmental impacts of protection may necessitate this choice. In the extreme case, an entire area, including building and transportation infrastructure, may be abandoned.
- Accommodation implies that people continue to use the land at risk but do not attempt to prevent the land from being flooded. This option includes erecting emergency flood shelters, elevating buildings on piles, converting agriculture to fish farming, and growing flood or salt tolerant crops.

- Protection involves hard structures such as sea walls and dikes, as well as soft solutions such as dunes and vegetation, to protect the land from the sea so that existing land uses can continue.[4]

Figure 6-3. Decadal Average Hurricane Total Dissipated Energy (in terms of PDI), Loss of Life, and Inflation-Adjusted Economic Damages ($000) from Hurricanes Making Landfall in Continental United States

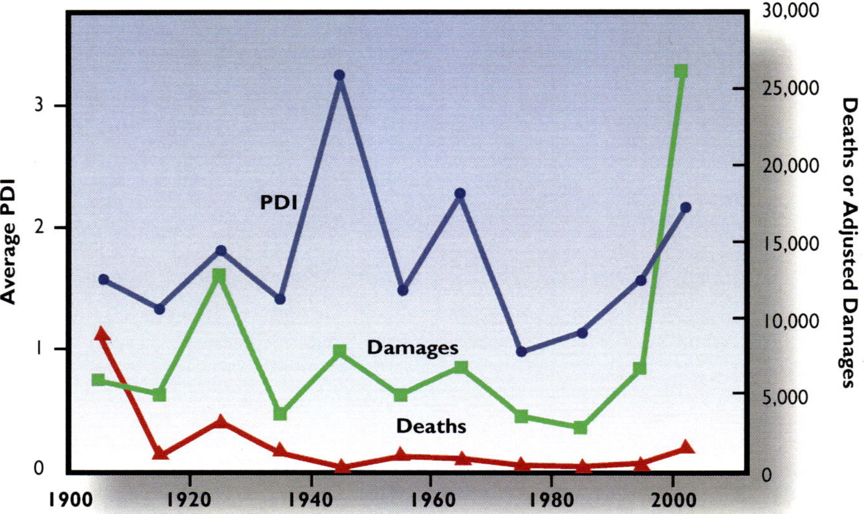

> Small islands, whether located in the tropics or higher latitudes, have characteristics which make them especially vulnerable to the effects of climate change, sea level rise, and extreme events.[2]

Federal leaders with facilities or stakeholders in coastal areas should prepare for changes predicted with high confidence by the scenarios and model results from the IPCC AR4. For facilities, this may involve relocation, taking steps to reduce their vulnerability to storm and wave damage and flooding, and preparing emergency plans. To address changing relations with stakeholders in light of anticipated changes in coastal operations, the federal leader may need to reevaluate services, needs, and capabilities. Preparation for extreme events in coastal settings improves with early warning systems, hazard awareness education, evacuation planning, and notification.[5]

Adaptation Example: Boston

Many of these adaptation issues are present in the construction of a sewage treatment plant in a coastal setting. Development and population growth in Boston, and the resulting pollution problems they bring, required construction of a new sewage treatment plant, and the best available location was on Deer Island in Boston Harbor. The low elevation allows gravity transport of the waste to the plant; such locations are invariably technically preferred. This was a major new infrastructure project investing in a vulnerable location (a small, coastal island of low elevation).

The likelihood of sea level rise during the lifetime of the facility presented two choices to the design engineers: future adaptation in response to sea level rise (reactive adaptation) or initial planning and construction in preparation for anticipated sea level rise (proactive adaptation). The reactive scenario would involve a protective wall around the plant and installation of expensive pumping equipment to transport the effluent over the wall. Model scenarios of climate change and sea level rise provided the basis for the proactive scenarios. Project managers chose the latter, and in 1998 the facility was constructed 2 feet higher than originally planned.

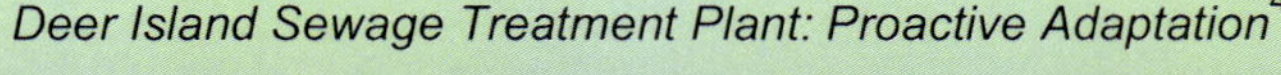

Deer Island Sewage Treatment Plant: Proactive Adaptation[4]

Coastal Management

With few exceptions, such as Deer Island, coastal management programs are not incorporating consideration of the potential consequences of climate change. Coastal management programs in Maine,[6] Rhode Island, South Carolina, and Massachusetts have implemented various forms of "rolling easement" policies to ensure that wetlands and beaches can migrate inland as sea level rises and coastal landowners and conservation agencies can purchase the required easements.[7]

In estuaries such as the Chesapeake Bay, restoration efforts have focused on the reduction of nutrient over-enrichment from rivers draining into the estuaries. However, efforts to reduce this enrichment will have to contend with multiple climate change-related problems such as sea level rise, increased winter-spring discharges but reduced summer runoff, warmer sea surface temperatures, and greater shoreline erosion. Stormwater management, urban stream restoration, and fisheries management are strategies that seem the strongest candidates for additional development.[8]

Preparation for coastal disasters and interruptions to agency operations differ from that for other coastal challenges such as increased sea surface temperature (leading to bleaching of corals, etc.) and surface water acidification (which can affect structures). Other challenges more closely mirror those now understood in coastal communities, but the long-term effects of climate change may compromise even the best-prepared community's infrastructures. Long-term research and site-specific identification of these challenges may be among the best preparation tools that an agency leader can apply.

Various tools are available to the federal leader to prepare for climate impacts on coastal areas;[9] they include indexes of vulnerability and tools to evaluate community vulnerability, monetary economic valuation and estimates of the socioeconomic and environmental effects of disasters, geographic information systems, mapping return periods of extreme events, scenarios about the future, flood simulators, integrated models, and coastal zone management.

In many cases, practices appropriate to prepare for climate change are the same as those for coastal disasters that can arise from other sources (coastal subsidence, tsunamis, severe storms, etc.). Long-term threats from climate change may dwarf the challenges from many of these other sources and require new adaptation measures.

Challenges to Water Resources

As discussed in Chapter 5, the water resources of the United States are a second area where impacts of climate change are highly likely and that already are experiencing increasing stress. Because the federal government manages large swaths of land, especially in the western United States, that are stressed already by limited water supplies, we discuss some adaptation measures currently called for or planned.

With reduced water supplies, federal leaders can also expect facilities to be in greater danger of damage from forest fires, and these fires may disrupt operations, transportation, and stakeholders. You may face demand to contribute resources during emergency events and can expect increased cost of operations in communities subject to forest fires.

Recognizing the large impacts that climate change can exert on the operations of public lands, the GAO has called for the Departments of Agriculture, Interior, and Commerce, which together manage over 90 percent of federal lands in the United States, to develop internal plans for dealing with the impacts of climate change upon their operations. GAO concludes that many agencies with responsibilities for public lands have not made sufficient progress on such planning and that resource managers in general are not receiving sufficient guidance in planning for climate change.[10]

Federal leaders must consider facilities and operations in western states that are vulnerable to impacts of climate change upon water supplies. Lack of water supplies to maintain facilities, operations, and needs of the staff can directly affect agency operations. Water shortages experienced by suppliers, clients, and other stakeholders will indirectly affect agencies. The magnitude of change possible may dwarf the experience of water managers, and extreme solutions may be required. These solutions may include alternative sources of water or relocating facilities and personnel.

The first step is to assess the range of forecasted impacts from various scenarios of climate change and consider the regional variability of those impacts in light of the distribution of facilities within the agency. Agencies should evaluate vulnerabilities within those parameters. The magnitude of adjustment that may be required and resources needed to make that adjustment may require significant resources not currently planned. Agencies may need long-range budget adjustments.

> There is very high confidence that adaptation procedures and risk management practices for the water sector are being developed in some countries and regions that have recognized projected hydrological changes with related uncertainties.[2]

Water utilities in several large U.S. metropolitan areas (such as Seattle, Miami-Dade, New York, and Southern California) now incorporate climate change information into long-term planning efforts. Because of the difficulty of detailed projections of climate change at the scale required for specific planning and because changes in climate are multifaceted, the key lesson now applied is the increased uncertainty climate change introduces into the planning effort. Variability in the future is likely to exceed that previously incorporated in planning,[11] and warmer temperatures are already exerting a significant hydrologic impact in the western United States. The American Association of State Climatologists recommends that policies focused on long-term climate change emphasize robust policy options for an enhanced range of alternatives not tied to specific predictions,[12] which are impossible to verify in the short term.

Experiences in the water sector in the United States and other countries can provide a meaningful and appropriate guide for developing adaptation procedures in other areas. Means for estimating the probability and magnitude of events, even from sparse data, are well developed in this field, as are methods of optimizing operations on the basis of that information. Continued development of models for downscaling global climate information can inform other spatially disaggregated estimates of the impacts of climate change.

Climate change is projected with very high confidence by mid-century to reduce water resources in many small islands, e.g., in the Caribbean and Pacific, to the point where they become insufficient to meet demand during low-rainfall periods.[2]

SHORT-TERM STEPS

The implications for federal agency operations, infrastructure, natural resource stewardship, and stakeholder engagement roles, although uncertain, can be significant.

By better understanding, planning for, and adapting to the climate changes that are confidently anticipated, federal agencies can take advantage of opportunities and reduce risks to their operations and programs. The process of understanding, planning, and adapting must be iterative because our knowledge and expectations are dynamic, and the major policy decisions made about mitigation will influence future challenges in adaptation.

The process of planning for such changes and continual improvement in the response to changes requires a systematic and well-founded approach. Federal programs have long used the process of adaptive management. It requires experimentation with alternative responses and evaluation of the results, continually testing how well you are doing, and changing your approach as you learn more.[13] Proactive adaptive management is not new to federal leaders as it serves as the framework for the management of federal organizations (continuous process improvement), infrastructure (strategic asset management), and natural resources (adaptive ecosystem management).

Some impacts will be favorable; others may have favorable and unfavorable results, and we expect all to change over time. Current, planned, and unanticipated treaties, laws, regulations, and EOs will circumscribe your ability to plan and execute adaptation strategies.

We will correctly anticipate some impacts, but given the complexity of the system and uncertainties in our projections, many of the impacts will be unanticipated. Address the former with "no regrets" policies (choose a policy that is helpful even if the impact does not occur, such as one with health co-benefits, say, reducing air pollution) and establish triggering events. The latter may require regular review of policies and actions.

Impacts are very likely to increase due to increased frequencies and intensities of some extreme weather events. Recent events have demonstrated the vulnerability of some sectors and regions, including in developed countries, to heat waves, tropical cyclones, floods, and drought, providing stronger reasons for concern as compared to the findings of the third assessment report.[2]

Impacts of climate change will become more severe and extensive as the magnitude of climate change increases, which will depend, in part, on mitigation plans implemented in the next decade. Therefore, a part of the agency leader's task will be to understand alternative scenarios and resulting projected trajectories of climate change (see Chapter 2) and prepare impact assessments and adaptation strategies for various planning horizons, which may extend well beyond typical agency planning horizons.

Decision as to the adaptation policies to employ depends on (1) the severity of the impact, (2) the risk of the impact, and (3) the ability of your organization to adapt. The concept of risk, which combines the magnitude of the impact with the probability of its occurrence, captures uncertainty in the underlying processes of climate change, exposure, impacts, and adaptation.[14]

Each of these (impact, risk, and adaptive capacity) varies dramatically across your agency's geographic and organizational areas of responsibility. Knowledge, financial resources, and institutional barriers define adaptive capacity, so part of your job will be to evaluate your adaptive capacity and take steps to increase it as agency vulnerability demands. Significant resources in the government and elsewhere can facilitate your adaptation decisions.

Federal leaders can do the following to prepare their agencies to adapt to climate changes:

- Adopt a proactive climate change adaptation framework within your agency strategic planning framework.
- Conduct a baseline assessment of key conditions.
- Engage stakeholders.
- Assess impacts and vulnerabilities.
- Assess adaptation options to mitigate vulnerabilities.
- Select options and develop budget requests and implement changes.
- Monitor and track progress.
- Review and refine strategies, policies, and plans.

The following sections describe these steps.

Adopt a Climate Change Adaptation Framework

Laws, EOs, and court orders, as well as policies and regulations, will guide much of your agency's response to climate change. However, you and your management staff can implement a few key actions regardless of the nature and scope of government-wide policies and regulations. The first step is for federal leaders and their management staff members to integrate the process of assessing vulnerabilities, evaluating climate change impacts, and developing response actions into the agency strategic planning processes for

- internal operations and supply chain management;
- infrastructure management (such as site options, design standards, operations and maintenance, and expected useful life);
- natural resources management (such as national parks, national forests, wildlife refuges, water resources, air quality, marine protected areas, and national estuaries); and

- stakeholder engagement and outreach. You will also be called upon to coordinate with and support other agencies on many of these issues.

This effort, albeit a formality, establishes clear leadership intent for assessing, preparing for, and appropriately responding to future climate conditions.

Conduct a Baseline Assessment

A meaningful assessment of the impacts of climate change requires a thorough understanding of the key relevant conditions, including the following:

- *Current agency conditions.* Effective adaptive planning relies on clear objectives or performance indicators of the agency conditions that must be maintained in the face of changing climate conditions, including
  - operational conditions (process cycle time, cost per unit, quality, etc.);
  - facility and infrastructure conditions (utilization, condition, energy use, etc.);
  - employee and constituent well-being conditions (satisfaction, injury or illness rates, etc.); and
  - supply chain conditions (supplier availability and transportation costs).
- *Baseline climate conditions.* Specification of the present-day or baseline climate is just as important as the specification of the future scenarios of climate change.

Baseline climate information is important for characterizing the prevailing conditions under which a particular agency functions and to which it must adapt. Such a baseline includes a description of average conditions, spatial and temporal variability, and anomalous events, some of which can cause significant impacts. Baselines also allow calibrating and testing impact models across the current range of variability, identifying possible ongoing trends or cycles, and specifying the reference situation with which to compare future changes. Agencies should consider the cost of obtaining this baseline condition data as part of its overall cost evaluation process. When possible, impact and adaptation assessments should use the most recent 30-year climate period as the climatological baseline period (*climate normal*).

The 1961–1990 normal period is frequently selected as the standard reference for many climate change studies, as it is considered to represent the present-day or recent average climate in the study region. The 30-year period is also sufficient duration to encompass a range of climatic variations, including a number of significant weather anomalies. In this period, data on all major climatological variables are abundant, adequately distributed over space, and readily available. Using modern records for the "normal period" provides data of sufficiently high quality for use in evaluating impacts and is consistent or readily comparable with baseline climatologies used in other impact assessments.

Assess Impacts and Vulnerabilities

Preparing for proactive adaptation will require you to assess impacts that may not be obvious, but thorough planning with lead-time can reduce the impacts and cost. We can observe impacts (so the adaptation is reactive), forecast impacts from current trends, or predict impacts by models using the results of climate change scenarios as described in Chapter 2. A good starting point to understand the impacts you will encounter is the IPCC tabulation presented in this book (see purple boxes throughout this book and Appendix C); in particular, the impacts it judges to have a likelihood of at least 90 percent (see Chapter 1).

> ***Vulnerability** to climate change is the degree to which systems are susceptible to, and unable to cope with, adverse impacts.*

In exercising due diligence, federal agency leaders can be expected to address, at a minimum, changes identified or projected with high confidence and those projected to be very likely or virtually certain. The impacts can operate in two directions: some happen to your agency and others your agency creates (as part of that cascading chain of events). You will be expected to identify which of your programs create impacts, what systems under your responsibility are particularly vulnerable, your agency's ability to adapt to these changes, and the cost-effective actions that will reduce the vulnerability of the systems. This will require identification of relevant stakeholders, development of risk management strategies, and allocation of resources to selected adaptation strategies. Your agency's normal planning horizons may be called into question, and you may decide to expand the time frame for strategic planning beyond 25 years.

Although we can adapt human systems and modify engineered systems, perhaps with more difficulty and delay, natural systems in your charge may lack the ability to adapt at rates required to maintain stability in the coming decades. When you have charge of public lands—treed strips surrounding your buildings, parks, staging grounds, distribution areas, or national parks—assessment of vulnerabilities (and cascading risks to other parts of your system) must recognize these increased risks.

Elements of your systems will vary in their *exposure* to climate change and in their *sensitivity*. For example, people living in Alaska may be more exposed to increased temperatures although the sensitivity to increased temperature may be greater in mid-latitude cities. Older people tend to be more sensitive to the effects of increased temperature than those in their 20s. Assessments of vulnerabilities must consider both exposure and sensitivity. As another example, people living in tropical areas may be more exposed to malaria and those with prior weakness may be more sensitive to the effects. We can reduce our vulnerability to malaria by use of mosquito nets (reducing exposure) among weakened (more sensitive) populations.

The complexity of changes, including spatial and temporal differences means some sectors and areas will be more stressed than others, and the uneven distribution of resources means the ability to adapt will vary. Therefore, effective management and sharing of resources will challenge the federal response.

Your agency's *vulnerability assessment* will require input from your stakeholders, historical data, and experience, as well as forecasts of future social and economic conditions and climate scenarios.[15] Table 6-2 shows the steps of an accepted method.

Table 6-2. Steps in Vulnerability Assessment

1. Engage affected parties	Engage and retain decision makers and those affected by future climate change.
2. Assess current vulnerability	Use experience to assess impact and potential damage. Understanding adaptive capacity, critical thresholds, and coping ranges is helpful.
3. Estimate future conditions	Use climate, environmental, and socioeconomic scenarios to determine future policy and development.
4. Estimate future vulnerability and identify adaptation strategies	Use the two previous steps (current vulnerability and future conditions) to identify future vulnerability and adaptation strategies.
5. Decide and implement	Incorporate results into risk-management strategies and follow through with them.

The IPCC uses seven criteria to determine whether vulnerabilities are *key*, a decision that helps policymakers determine whether they represent a dangerous anthropogenic influence in the sense of the UNFCCC, Article II. You should be more likely to consider these vulnerabilities as key if they have impacts characterized by the following:

- Large magnitude
- Imminent occurrence
- Persistence and irreversibility
- High likelihood
- Less potential for adaptation
- Heterogeneous distribution
- Risk to socially or economically valued systems.

Assess Adaptation Options

Choose adaptation options on the basis of a careful assessment of efficacy, risks, and costs. Due to uncertainties in projected climate changes and in how systems will respond to them, adaptation options carry varying degrees of uncertainty, or risk, as well. Timing, priority setting, economic and political costs, availability of resources and skills, and the efficacy of various solutions all should be a part of the discussion.

The Pew Center presents a range of approaches for assessing adaptation options,[16] including the following:

- *No-regret.* Actions that make sense or are worthwhile regardless of additional or exacerbated impacts from climate change, for example, protecting or restoring systems that are already vulnerable or of urgent concern for other reasons.
- *Profit or opportunity.* Actions that capitalize on observed or projected climatic changes, for example, a farmer is able to shift to different crops better suited to changing climatic conditions.
- *Win-win.* Actions that provide adaptation benefits while meeting other social, environmental, or economic objectives, including climate change mitigation, for example, improving the cooling capacity of buildings through improved shading or other low-energy cooling solutions.

- *Low-regret.* Measures with relatively low costs for which benefits under climate change scenarios are high, for example, incorporating climate change into forestry, water, and other public land management practices and policies or long-term capital investment planning.
- *Avoiding unsustainable investments.* Policies or other measures that prevent new investment in areas already at high risk from current climatic events, where climate change is projected to exacerbate the impacts, for example, prohibiting new development in flood-prone areas where sea level rise is increasing and protective measures are not cost effective.
- *Averting catastrophic risk.* Policies or measures intended to avert potential or eventual catastrophic events, or those so severe or intolerable that they require action in advance on the basis of available risk-assessment information, for example, relocating Alaskan villages in areas at or near sea level with projected sea level rise and increasing severe weather events.

Regardless of approach, leaders will have to assess the cost and benefits of the alternatives. At present, reliable information is lacking on the costs of climate impacts, making it difficult for decision makers to judge the amount of resources that they should allocate to adaptation.

Federal agencies should apply a standard method for costing climate impacts and comparing them with the costs of adaptation measures. The method should enable decision makers to calculate valid, order-of-magnitude estimates of the costs to help identify priority climate risks and select appropriate adaptation measures. The process for identifying costs and benefits should consist of the following four steps:

1. Identify and measure (quantify) climate impacts in physical units—the direct impacts of climate change and indirect effects associated with implications of the direct impacts.
2. Convert these physical impacts into monetary values.
3. Calculate the resource costs of the proposed adaptation options.
4. Assess the costs and benefits of the adaptation options and choose the preferred options, taking into account the risks and uncertainties.[17]

Step 1 can include the number of buildings damaged, miles of roads that need repair, or people without a home. Some of these impacts can be easy to quantify and some may be heart wrenching. Some impacts are intangible, such as the historical value of an item or its cultural importance. Here the federal leader will also face political pressures that reveal the limitations of strict cost-benefit analysis.

In undertaking Step 4, one could use cost-benefit analysis to determine whether the total benefits of an adaptation option (or suite of options) are greater than the costs. Cost-benefit analysis can complement other decision-support tools, such as multicriteria analysis or Pareto optimal solutions, in cases where criteria other than economic efficiency are important.

In identifying the costs and benefits of climate change impacts and adaptation, the costing method that an agency chooses must articulate assumptions about the causation of socioeconomic changes. This allows the user to identify the potential effects, which are solely due to climate change.

Select and Implement Changes

Once climate impacts have been valued, and the resource costs of the various adaptation options calculated, the decision maker needs to bring this information together and engage appropriate stakeholders to compare the outcomes of each adaptation option and identify the preferred course of action.

A decision-making body can use the following decision-support tools to help select the preferred option:

- When outcomes are expressed in economic terms, evaluation of alternatives may be performed using a cost-benefit analysis, which demonstrates whether the total benefits of an adaptation option are greater than its costs.
- However, economic value will seldom be the sole criterion for decision making—other objectives are likely to be important, too. In these cases, other decision-support tools, such as multicriteria analysis, account for these wider considerations.
- Various selection criteria differentiate between options, depending on the quality of the decision maker's knowledge. When knowledge of the probability of an event is poor (as will often be the case with climate change), criteria such as "maximin" or "minimax regret" can be used. Other techniques such as "net present value" or "expected net present value" are

useful when the decision maker has greater certainty about outcomes.

- The agency's decision-making body will want to know the sensitivity of the estimates to the input data and models used in the analysis. They will also need to understand any key assumptions. Techniques for testing the factors that underpin the estimated outcomes include sensitivity analysis, simulation, and interval analysis.[17]

A discussion of the decision-support tool applied and the outcome of the analysis, together with supporting documentation on selected adaptation options, must be included in the agency budget documents to secure the appropriate funding.

Measure Conditions and Evaluate Progress

Incorporating adaptation to climate change into an agency's routine strategic planning is a challenge, particularly because most agencies set a wide range of strategies affecting many levels of the organization.[18] An interdisciplinary, cross-functional approach is crucial in adaptation planning.

> ***Monitoring*** *is a process of tracking progress toward a goal or objective, specifically the implementation of an adaptation strategy and its components.*[18] ***Evaluation*** *is "a process for systematically and objectively determining the relevance, efficiency, effectiveness, and impact of an adaptation strategy in light of its objectives."*[18]

To measure progress and ensure integration, adaptation planning needs to include monitoring and evaluation. Defining appropriate metrics, including goals, objectives, indicators, outputs, and outcomes, gives the staff the underlying framework needed to evaluate progress toward adaptation goals. Through monitoring, the staff is able to take timely corrective action and measure success through evaluation. A useful mnemonic for remembering your objectives is to make them S-M-A-R-T: specific, measurable, achievable, relevant, and time bound.[19]

Agency leaders may ask the staff to conduct self-assessments or arrange external evaluation of performance. Current measurement and evaluation teams may be able to incorporate climate change concerns into established processes.[18]

United Nations Development Programme's report, *Continuing the Adaptation Process*, characterizes participatory monitoring and evaluation programs, where the most vulnerable groups are included in planning, as "the trend" in the field of climate adaptation.[18] These groups can be included in a network that participates in measurement, monitoring, evaluation, and decision making.

To achieve successful monitoring and evaluation, the federal manager must ensure meaningful data are collected and maintained continuously throughout a project. Project managers can monitor progress using the in-place management information system. Those managing a project or program day to day should be involved in monitoring progress toward its objectives and goals (for example, the facility management staff should monitor the performance of buildings with new design features to reduce energy). Thus, appropriate monitoring and evaluation metrics and processes may differ greatly at each level of the organization.

Monitoring and Evaluation Metrics and Examples

An agency's facilities are threatened by sea level rise. A strategy to adapt to sea level rise might include the following metrics.

Goal. *The desired overarching result of a specific strategy. Example: Protect the agency from sea level rise.*

Objective. *A specific, measurable, planned incremental result to be achieved in a specific time frame. Example: Reduce the percentage of facility space at risk of flooding from sea level rise by 50 percent by 2014.*

Indicator. *Data collected to measure intended progress toward an objective. Example: The percentage of facility space at risk of flooding from sea level rise.*

Output. *A measurable, immediate, and intended result of the project. Example: The number of facilities protected from flooding by a sea wall.*

Outcome. *The overarching result of a strategy. Example: The agency is protected from sea level rise.*

A monitoring and evaluation program provides the basis for analyzing the impact and outcome of an intervention by comparing data collected at the baseline with those collected over the course of a project or at its conclusion. The analysis may uncover both positive and negative effects and unanticipated consequences.[18]

Each adaptation strategy should be reviewed to determine the information the staff will need to evaluate progress toward the goals of that strategy. As the agency implements its strategies and makes progress, the frequency of monitoring and metrics used may change, but the practice of monitoring and evaluation must continue.

Understand Adaptation and Mitigation

Decisions about adaptation may influence and be influenced by *mitigation* actions (see Chapter 4) taken on a national or global scale outside of your direct responsibility. For example, protecting shorelines by wetlands restoration (increasing adaptation) may increase the release of the GHG methane (reducing mitigation). Reforestation efforts (mitigation through carbon sequestration) could reduce the availability of resources needed for adaptation. External factors could affect both adaptation and mitigation. For example, melting permafrost will disrupt transportation of items needed for adaptation and release GHGs, thereby stymieing efforts at mitigation. Therefore, broad stakeholder involvement and interagency consultation, as well as impending legislative changes in mitigation strategies, should inform your efforts.

Although we may think of adaptation as one means to "buy time" for mitigation efforts to take effect, there is no clear evidence this will work. However, building the *capacity* to adapt is very likely to improve our ability to mitigate climate change.[20]

Adaptation requires financial, natural, and knowledge resources, and earlier response ensures those resources are available. For example, natural systems generally respond to change negatively, more so if the stress is rapid or large, so effective mitigation improves successful adaptation. Part of your job as a federal leader may be to articulate the need for mitigation so that your efforts at adaptation can succeed.

WHERE TO LEARN MORE

To learn more about adaptation to climate change, you should review the sources that apply to your agency:

- IPCC resource
  - IPCC, *Climate Change 2007: Impacts, Adaptation and Vulnerability, Working Group II Contribution to the Fourth Assessment Report of the Intergovernmental Panel on Climate Change*, www.ipcc.ch/ipccreports/ar4-wg2.htm
- Federal government agency resources
  - U.S. CCSP, Goal 4 Synthesis and Assessment Reports 4-1 to 4-7, www.climatescience.gov/Library/sap/sap-summary.php
  - U.S. EPA

        - "State Climate Action Plans Database," *Climate Change—What You Can Do*, yosemite.epa.gov/gw/StatePolicyActions.nsf/webpages/index.html
        - "State Action Plan Recommendations Matrix," *Climate Change—What You Can Do*, yosemite.epa.gov/gw/StatePolicyActions.nsf/matrices/0?opendocument
    - U.S. Environmental Protection Agency, "Local Resources," *Climate Change—State and Local Governments*, www.epa.gov/climatechange/wycd/stateandlocalgov/local_resources.html
- State and local resources
    - ICLEI, Local Governments for Sustainability, "Cities for Climate Protection (CCP)," *Programs*, www.iclei.org/index.php?id=800
    - California Energy Commission, "The Energy Yardstick: Using PLACE^{3}S to Create More Sustainable Communities," *PLACE^{3}S*, www.energy.ca.gov/places/
- Nongovernmental organization and business resources
    - Pew Center on Global Climate Change, *Business*, www.pewclimate.org/business
    - World Resources Institute, *Publications: Climate, Energy, & Transport*, www.wri.org/publications/climate
    - Heinz Center, *A Survey of Climate Change Adaptation Planning*, www.heinzctr.org/Press_Releases/adaptation_survey.shtml

---

[1] IPCC Fourth Assessment Report sea level rise does not include effects of land subsidence and uplift or land-based ice melt.

[2] IPCC key finding. Throughout this book, each purple box shows a key finding from the IPCC with a high level of scientific understanding. See Appendix C for specific AR4 chapter sources for each finding. Chapter 1 gives a full discussion of our selection of key findings.

[3] The power dissipation index is the integral of the cubed maximum sustained wind speed over all 6-hour observations when a storm is at least at tropical storm intensity (18 m/s), as described in Kerry Emanuel, "Increasing

Destructiveness of Tropical Cyclones over the Past 30 Years," *Nature*, Vol. 436, No. 7051 (August 4, 2005), pp. 686–688, ISSN 0028-0836.

[4] Coastal Zone Management Subgroup, IPCC Response Strategies Working Group, *Strategies for Adaption to Sea Level Rise*, November 1990, www.epa.gov/climatechange/effects/coastal/SLRAdaption.html.

[5] Source for photo: Massachusetts Water Resources Authority, The Deer Island Sewage Treatment Plant, www.mwra.state.ma.us/03sewer/html/sewditp.htm.

[6] EPA, Policy, Planning, and Evaluation and Maine State Planning Office, *Anticipatory Planning For Sea-Level Rise Along The Coast of Maine*, EPA-230-R-95-900, September 1995, www.epa.gov/climatechange/effects/downloads/maine_0.pdf.

[7] EPA, "Coastal Zones and Sea Level Rise," *Climate Change—Health and Environmental Effects*, August 19, 2008, www.epa.gov/climatechange/effects/coastal/index.html.

[8] Christopher R. Pyke, Raymond Najjar, et al., *Climate Change and the Chesapeake Bay: State-of-the-Science Review and Recommendations,* A Report from the Chesapeake Bay Program Science and Technical Advisory Committee, April 21, 2008, www.chesapeake.org/stac/QuarterlyMeetingPresentations Material/june08/climatechangefinaldraft.pdf.

[9] IPCC, "Coastal Systems and Low-Lying Areas" in *Climate Change 2007: Impacts, Adaptation, and Vulnerability. Working Group II Contribution to the Fourth Assessment Report of the Intergovernmental Panel on Climate Change*, Table 6.10, Cambridge University Press, Cambridge, United Kingdom, and New York, NY, USA, www.ipcc.ch/pdf/assessment-report/ar4/wg2/ar4-wg2-chapter6.pdf.

[10] John B. Stephenson et al., *Climate Change: Agencies Should Develop Guidance for Addressing the Effects on Federal Land and Water Resources*, Government Accountability Office (GAO)-07-863 (Washington, DC: GAO, August 2007), www.gao.gov/new.items/d07863.pdf.

[11] K. Miller and D. Yates, *Climate Change and Water Resources: A Primer for Municipal Water Providers*, American Water Works Association and University Corporation for Atmospheric Research, Project #2973, 2006, p. 83.

[12] American Association of State Climatologists, *Policy Statement on Climate Variability and Change*, 2002, www.stateclimate.org/publications/files/aascclimatepolicy.pdf.

[13] CCSP, 2008, *Preliminary Review of Adaptation Options for Climate-Sensitive Ecosystems and Resources*, A Report by the U.S. Climate Change Science Program and the Subcommittee on Global Change Research, S. H. Julius and J.M. West, eds. (Washington, DC: USEPA, 2008), 873 pp., www.climatescience.gov/Library/sap/sap4-4/final-report/.

[14] IPCC, "Assessing Key Vulnerabilities and the Risk from Climate Change" in *Climate Change 2007: Impacts, Adaptation, and Vulnerability. Working Group II Contribution to the Fourth Assessment Report of the Intergovernmental Panel on Climate Change*, Chapter 19, Cambridge University Press, Cambridge, United Kingdom, and New York, NY, USA, www.ipcc.ch/pdf/assessment-report/ar4/wg2/ar4-wg2-chapter19.pdf.

[15] D. S. Lemmen and F. J. Warren, eds., *Climate Change Impacts and Adaptation: A Canadian Perspective* (Natural Resources Canada, 2004), 174 pp., adaptation.nrcan.gc.ca/perspective/pdf/report_e.pdf.

[16] Pew Center on Global Climate Change and the Pew Center on the States, "Climate Change 101: Adaptation" part of the *Climate Change 101: Understanding and Responding to Global Climate Change* series, www.pewclimate.org/docUploads/Adaptation_0.pdf.

[17] Metroeconomica Limited, *Costing the Impacts of Climate Change in the UK: Implementation Guidelines*, Final Report for The UK Climate Impacts Programme, June 18, 2004, www.ukcip.org.uk/images/stories/Pub_pdfs/CostingsImplementationGuidelines.pdf.

[18] Rosa T. Perez, et al., "Continuing the Adaptation Process," in *Policy Frameworks for Climate Change: Developing Strategies, Policies, and Measures*, United Nations Development Program, and Global Environmental Facility, www.undp.org/gef/undp-gef_publications/publications/apf%20technical%20paper09.pdf.

[19] U.S. Department of State (DOS), Bureau of Population, Refugees and Migration (PRM), *Writing Goals, Objectives, and Indicators for PRM-funded Humanitarian Assistance Projects*, www.state.gov/g/prm/109569.htm.

[20] IPCC, "Inter-relationships between Adaptation and Mitigation" in *Climate Change 2007: Impacts, Adaptation, and Vulnerability. Working Group II Contribution to the Fourth Assessment Report of the Intergovernmental Panel on Climate Change*, Cambridge University Press, Cambridge, United Kingdom, and New York, NY, USA, www.ipcc.ch/pdf/assessment-report/ar4/wg2/ar4-wg2-chapter18.pdf. [See especially Table 18.1.]

Chapter 7
International Climate Change Response

Loss of biodiversity in Australia and New Zealand, species loss in Europe, reduced water quantity in the Caribbean, flooding in South Asia, and drought in sub-Saharan Africa: climate change is international. Directly and indirectly, these changes in other countries will affect the United States and operations of the U.S. government. Therefore, federal leaders must understand international climate policy and the range of potential climate change in other countries. This chapter addresses two questions:

1. What opportunities and challenges does international climate change policy present to my agency?
2. How could climate change outside the United States involve my agency?

The international community's diplomatic work to address climate change effects holds increasing importance for the United States. International acceptance of the reality and challenge of climate change to individual countries and pressure to mitigate this change by reducing GHG emissions are increasing. Climate change also affects shared global resources—oceans, marine fisheries, the atmosphere, Arctic ice shelves, and Antarctic ice sheets—and these in turn influence the U.S. economy, food supply, health, and other societal concerns. Only international agreements and action can address a problem with a shared source that affects these shared resources. We cannot solve the challenge solely within the borders of one country. It requires international cooperation and creates winners and losers among countries.

Climate change in individual countries produces ripple effects worldwide. Physical stresses of climate change on other nations will require U.S. resources, time, attention, coordination, and consultation. For example, changes in growing seasons or fresh water scarcity overseas may have indirect consequences for the federal government such as through changes in our food supply. The U.S. government will operate relief operations in response to climate change. Climate change can disrupt the supply of imported materials on which agencies currently rely to carry out their missions. In addition, multiple departments and agencies—DoD, DOS, USAID, and others—have significant operations overseas. International climate change could potentially disrupt the ability to support those operations. All of these interactions and complications lie in a framework of an increasingly globalized and interdependent economy and growing flow of people, capital, and information around the world.

Consistent with our approach in other chapters, we highlight the most significant findings related to international issues and discuss their implications for the federal manager. As discussed in Chapter 5, climate change affects virtually every economic and governmental sector. Just as the natural environment does not have institutional borders, impacts to the environment cross international lines.

International Aspects of Climate Change

What Opportunities and Challenges Does International Climate Change Policy Present to My Agency?

The observed and potential impacts of climate change drive international policy developments. Those policy developments in turn modify how climate changes and affects countries. This feedback promises to confound our attempts to deal with climate change.

Although we cannot predict the components of any international agreement, we expect it will increase U.S. government responsibility for reducing emissions in coming years.

If the United States commits to any type of international carbon agreement, the administration is likely to require agencies to reduce their GHG emissions. Your agency may participate in international forums and partnerships to manage mitigation internally and externally. Your agency will also have opportunities to address the impacts of climate change. This section presents international policy discussions, initiatives, and changes that may involve you. We briefly explain how the United States evolved to its current position and how this position is likely to change in the near future.

Although the issue of climate change has relatively recently reached public consciousness, evidence in the 1960s and 1970s revealed that concentrations of CO_2 in the atmosphere were rising. These findings led climatologists and others to push for action.[1] As a response to this growing international concern for climate change, the World Meteorological Organization and United Nations Environment Program established the

IPCC in 1988. The initial findings of the IPCC prompted creation of the UNFCCC, approved at the Earth Summit in Rio de Janeiro in 1992. The United States is one of 192 countries that have ratified the UNFCCC to date. As part of the convention, a conference of the parties (COP) is held annually. The COP relies on the findings of the IPCC to frame policy recommendations.

Under the UNFCCC, governments

- *gather and share information on GHG emissions, national policies, and best practices;*
- *launch national strategies for addressing GHG emissions and adapting to expected impacts, including the provision of financial and technological support to developing countries; and*
- *cooperate in preparing for adaptation to the impacts of climate change.*

One of the most notable tools of the COP is the Kyoto Protocol, which the third COP (COP 3) adopted in Kyoto, Japan, on December 11, 1997. Kyoto commits developed countries to stabilize emissions, whereas the UNFCCC simply encouraged them to do so. Following the adoption of the Kyoto Protocol, COP 7 adopted detailed rules for its implementation, called the "Marrakesh Accords," 4 years later.

The Kyoto Protocol entered into force in 2005, when a sufficient number of countries had ratified it. Although the United States is a signatory to the UNFCCC and supports the work of the IPCC (being the largest single funder of the activities of each), it has declined to ratify the Kyoto Protocol. One stumbling block is the fact that the protocol does not require GHG stabilization from all countries.[2] A second concern is the possibility it will reduce economic growth.

There is high agreement and much evidence that notable achievements of the UNFCCC and its Kyoto Protocol are the establishment of a global response to the climate change problem, stimulation of an array of national policies, the creation of an international carbon market, and the establishment of new institutional mechanisms that may provide the foundation for future mitigation efforts. Progress has also been made in addressing adaptation within the UNFCCC and additional initiatives have been suggested.[3]

Copenhagen will host COP 15 in December 2009. This meeting is a critical milestone for negotiating a post-Kyoto accord because the first GHG reduction commitment period of the Kyoto protocol ends in 2012.

The United States currently participates in 15 bilateral climate change agreements with countries and regional organizations—Australia, Brazil,

Canada, China, Central America (Belize, Costa Rica, El Salvador, Guatemala, Honduras, Nicaragua, and Panama), EU, India, Italy, Japan, Mexico, New Zealand, the Republic of Korea, the Russian Federation, and South Africa—cumulatively representing more than 70 percent of GHG emissions.[4,5]

The partnerships range in focus from science and technology to policy development. Each partnership involves multiple U.S. federal agencies. For example, the U.S.–Italy Climate Change Partnership, initiated in 2001, brings together government, academic, and private-sector representatives from each country annually. Representatives from DOE, EPA, USDA, DOS, NOAA, USAID, NASA, and CCSP attended the 2007 meeting. With hundreds of climate change-related initiatives and multiple countries, federal agencies will participate significantly in international climate change diplomacy beyond their participation in the multilateral UNFCCC.

> The literature provides high agreement and much evidence of many options for achieving reductions of global GHG emissions at the international level through cooperation. It also suggests that successful agreements are environmentally effective, cost-effective, incorporate distributional considerations and equity, and are institutionally feasible.[3]

Several key working groups are likely to affect future U.S. positions on climate change:

- *Globe +5 Climate Change Dialogue.*[6] G8 members (the United States, Germany, France, Canada, Italy, Japan, Russia, and the United Kingdom) and Brazil, China, India, Mexico, and South Africa. Develops specific regulatory and policy recommendations for legislators and governments.
- *The Asia Pacific Partnership on Clean Development and Climate.*[7] Australia, Canada, China, India, Japan, the Republic of Korea, and the United States. Accelerates the development and deployment of clean energy technologies.
- *The Carbon Sequestration Leadership Forum.*[8] Twenty-two member countries support cost-effective technologies for the separation and capture of CO_2 for its transport and long-term safe storage.
- *Major Economies Meeting on Energy Security and Climate Change.*[9] Sixteen member countries and the EU develop and

contribute to a post-Kyoto framework on energy security and climate change by the end of 2008.

The existence of these groups and U.S. support for them show how seriously many countries take this issue. Although the United States does not currently mandate national carbon management or reporting (see Chapter 3 for details of local, state, and regional programs), approaches of other nations may become a model for future U.S. programs.

Preparations for COP 15

Evidence indicates the next iteration of the Kyoto Protocol will call for action to stabilize GHG emissions on the basis of the most recent findings of the IPCC. As the COP renegotiates the Kyoto Protocol in 2009, the United States may ratify its successor.

In the Major Economies Meetings on Energy Security and Climate Change, the United States convened the 17 major world economies—those responsible for most of the world's energy use and GHG emissions—and the United Nations. Representatives at these meetings have begun to negotiate a post-2012 framework under the UNFCCC. Progress from these meetings provides one foundation for an agreement at COP 15.

> There is high agreement and much evidence that: New energy infrastructure investments in developing countries, upgrades of energy infrastructure in industrialized countries, and policies that promote energy security, can, in many cases, create opportunities to achieve GHG emission reductions compared to baseline scenarios. Additional co-benefits are country-specific but often include air pollution abatement, balance of trade improvement, provision of modern energy services to rural areas and employment.[3]

The United States is more likely to support an agreement, or aspects of an agreement, that encourages technology development in the energy sector. Provisions in the next international agreement that promote development of carbon sequestration technologies may be attractive to the United States.

Several U.S. federal agencies may have a direct role stemming from any U.S. commitments to pursue or support energy technology development. The DOE, EPA, DOI, and NASA may all have increasing roles in funding and managing alternative and advanced energy projects.

For the United States, an important issue in negotiating a new treaty is the year chosen for the benchmark reductions. Because some countries have been working since 1990 to slow GHGs, they would be disadvantaged if a new start date were enacted. However, if the 1990 date remains, the reductions required for the United States would be greater than those for later benchmark years. Our emissions are now growing at 3 percent per year (see Chapter 1).

Environmental Impacts around the World

As in the United States, the majority of the world's urban populations are near the coast, so the effects of climate change on other countries are likely to be as severe internationally as in the United States. Table 7-1 illustrates how some agencies may change their adaptive response as they manage the response to climate change in other countries.

Table 7-1. Federal Agency Adaptive Response to Climate Change

Agency	Adaptive response
NOAA	Protection of marine and coastal resources
USAID, OFDA	Humanitarian relief and disaster response
DoD	Immediate disaster response
DOS	Communication of U.S. policy and response coordination
DOE	Support of and involvement in development of advanced energy and carbon solutions
EPA	Potential consultation for and enforcement of mitigation efforts
USDA	Downstream impacts to U.S. and foreign agricultural sectors
FBI, CIA, DHS, DoD	Downstream impacts to national security
DOC	Impacts to foreign production and consumption of goods
DOT	Disruptions to air and marine transportation systems and infrastructure
DHS	Immigration and environmental refugee issues
CDC	International cooperation on disease-preventing agreements and initiatives

Note: OFDA = Office of Foreign Disaster Assistance; FBI = Federal Bureau of Investigation; CIA = Central Intelligence Agency; and CDC = Centers for Disease Control and Prevention.

Table 7-2 illustrates the varied coastal sectors that may experience the socio-economic repercussions of climate change. Temperature rise, extreme events, floods, and saltwater intrusion may have widespread effects on coastal communities worldwide.

Table 7-2. IPCC Summary of Climate-Related Impacts in Coastal Zones

Coastal socio-economic sector	Temperature rise (air and seawater)	Extreme events (storms, waves)	Floods (sea level, runoff)	Rising water tables (sea level)	Erosion (sea level, storms, waves)	Salt water intrusion (sea level, runoff)	Biological effects (all climate drivers)
Freshwater resources	+	+	+	+	–	+	x
Agriculture and forestry	+	+	+	+	–	+	x
Fisheries and aquaculture	+	+	x	–	x	+	+
Health	+	+	+	x	–	+	+
Recreation and tourism	+	+	x	–	+	–	+
Biodiversity	+	+	+	+	+	+	+
Settlements/infrastructure	+	+	+	+	+	+	–

Note: "+" = strong; "x" = weak; "–" = negligible or not established.

These changes could extend to shipping lanes and port operations and change the costs and lead-times for imported material as well as material exported to support government operations overseas.

> By the 2080s, many millions more people than today are projected to experience floods every year due to sea level rise. There is very high confidence that the numbers affected will be largest in the densely populated and low-lying megadeltas of Asia and Africa while small islands are especially vulnerable.[3]

Hurricane Katrina in Louisiana can be used to illustrate the consequences for coastal ecosystem services if such events increase in frequency and intensity. Five-hundred and sixty km^2 of coastal wetlands, levees, and islands—which served as a storm-surge buffer for New Orleans—were lost because of Hurricane Katrina.[10] In addition, these areas supported the area's strong commercial fisheries industry, and the loss of marshes and estuaries can have a major effect on the success of fish stocks.

On the international scene, Cyclone Nargis, the most devastating cyclone to strike Asia since 1991, illustrates these challenges. It struck Burma on May 2, 2008, with winds of more than 250 km/hr, producing severe flooding in low-lying delta areas. USAID estimates that 63,000–101,000 people died due to flooding and 2.4 million (one-third of the population in affected areas) were affected. Major rice-producing areas were flooded with saltwater, damaging crops and soils. Table 7-3 summarizes some of the involvement of U.S. agencies in recovery efforts.

Table 7-3. Agency Response to Cyclone Nargis

Agency	Operation
DOS	U.S. Embassy in Rangoon conducted assessments. State also provided humanitarian relief. Humanitarian Information Unit provided details about impacts.
USAID	OFDA led disaster relief and provided funds to UN World Food Program and American Red Cross. Coordinated public fundraising programs. Food For Peace program provided food supplies. Office of Transition Initiatives provided post-disaster transition assistance.
DoD	DoD Overseas Humanitarian, Disaster and Civic Aid, supported OFDA with military capabilities with Joint Task Force Caring Response headquartered in Khorat, Thailand. Completed 151 airlifts of emergency relief commodities from Utapao to Rangoon.
NASA	Multiple satellite platforms provided continuing coverage of the storm track, precipitation, winds, flooding, and damage.
NOAA	Satellite and Information Service provided rainfall data in cooperation with the Joint Typhoon Warning Center.
DOI	USGS provided satellite imagery to aid rescue and recovery efforts within hours of the initial request and continuing imagery to assess the extent of flood damage. The USGS provided imagery to other participating agencies under the *International Charter Space and Major Disasters*.
USDA	Foreign Agricultural Service analyzed extent of saltwater flooding damage to rice crops and production areas.
Treasury	Office of Foreign Asset Control issued General License No. 14 to allow certain financial transactions in support of humanitarian or religious activities by nongovernmental organizations in Burma.
Radio Free Asia and Voice of America	Alerted local populations to the impending cyclone and flooding.

Although attributing any one storm to climate change is impossible, both Hurricane Katrina and Cyclone Nargis provide examples of the extent of damage possible and help reveal the types of problems that the world will encounter if there are more events like these.

Australia and New Zealand—By 2020, significant loss of biodiversity is projected to occur with very high confidence in some ecologically rich sites, including the Great Barrier Reef and Queensland Wet Tropics. By 2050, ongoing coastal development and population growth in some areas of Australia and New Zealand are projected to exacerbate risks from sea level rise and increases in the severity and frequency of storms and coastal flooding.[3]

Figure 7-1 shows coastal deltas identified by the IPCC as vulnerable, as well as the size of the population at risk of being displaced by sea level rise by 2050. The average population of deltas is 500 people per km^2. Three coastal areas identified as having relatively extreme vulnerability include the Nile Delta in Egypt, Ganges Brahmaputra Delta in Bangladesh, and Mekong Delta in Vietnam. Many other areas face high or medium risk.[11] Cyclone Nargis struck the Irrawaddy Delta, one delta region vulnerable to climate change that was not identified in Figure 7-1, showing that additional deltas not specifically identified as being at high risk are subject to significant climate impacts.

Figure 7-1. Relative Vulnerability of Coastal Deltas

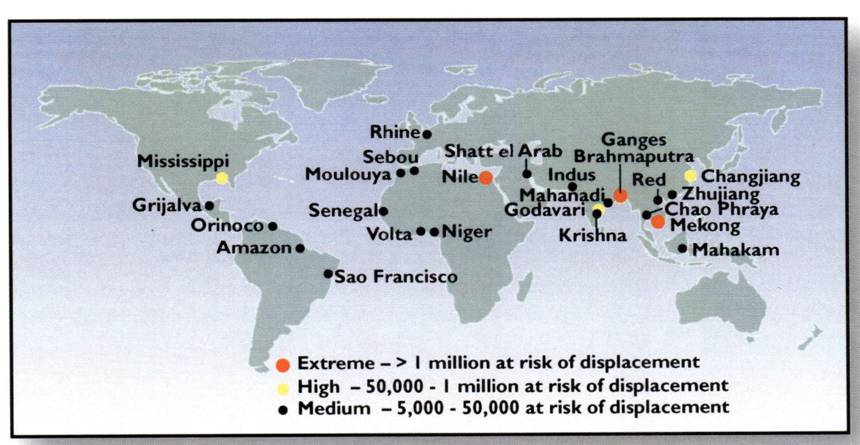

Climate change impacts to these areas could affect the internal stability of nations, which could in turn could have a significant impact on the United States. The National Intelligence Council explains:

> An impact [of climate change] is significant [to the U.S.] when it causes a noticeable—even if temporary—degradation in one of the elements of national power (geopolitical, military, economic, or social cohesion) because it indirectly influences the U.S. homeland, indirectly influences the United States through a major military ally or a major economic partner, or because the global impact is so large that the United States cannot be unaffected (i.e., it indirectly consumes U.S. resources).[12]

Cyclone Nargis illustrates how a natural disaster in a major delta can have international political complications. As these events grow in frequency, we may face complex events like Cyclone Nargis, where we must provide aid, but will also face the unique dynamics of the government of each affected country.

The impact of Cyclone Nargis illustrates a continuing challenge for major delta areas, which face the consequences of human-imposed change on the environment, coupled with future impacts of climate change. Southeast Asia is a large source of material for U.S. consumers, including government agencies. China, Japan, South Korea, Taiwan, and Malaysia alone represent about 30 percent of U.S. imports.[13] Climate change impacts in this region could impact the availability of these goods, potentially affecting the ability of agencies to carry out current business activities.

> Europe—There is very high confidence that negative impacts will include increased risk of inland flash floods and more frequent coastal flooding and increased erosion (due to storminess and sea level rise).[3]

Figure 7-2 shows the projected increases and decreases in global precipitation on the basis of SRES A1B scenario (see Chapter 2) for winter and summer.[14] These projections show the world becoming drier in some areas where access to water and drought are already impacting the population (for example, northern Africa and the southwestern United States), while other areas show increased precipitation. The strongest area of agreement of the models is large increases (over 20 percent) in Arctic regions (stippled gray).

These two examples, flooding and precipitation changes, illustrate the global character of climate change and the challenges of climate change facing the United States. Many more impacts are projected (see Chapter 5); each will have global ramifications and each is likely to impact the operations of many federal agencies.

> Europe—There is very high confidence that mountainous areas will face glacier retreat, reduced snow cover and winter tourism, and extensive species losses (in some areas up to 60 percent under high emissions scenarios by 2080).[3]

Figure 7-2. Projected Relative Percentage Changes in Precipitation by 2100

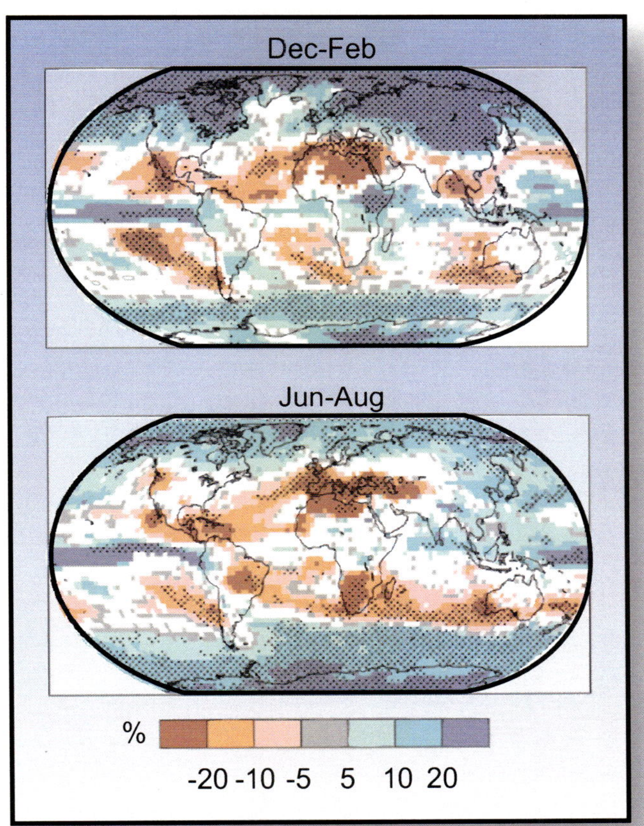

U.S. Involvement

How Could Climate Change Outside the United States Involve My Agency?

The CCSP supports research on international climate change (see Chapter 1) through internal programs and bilateral agreements. For example, the USAID Global Climate Change Program integrates climate change into the agency's development activities. The CCSP has published a variety of reports on climate change adaptation and mitigation. In addition, USAID has participated in bilateral partnerships on climate change with Brazil, Central America, India, Mexico, Russia, and South Africa.[15]

Increased involvement by other agencies is likely as climate change affects the world community. In addition, the downstream effects of climate change will affect many organizations indirectly. The involvement of the United States will vary on the basis of the countries affected. Shifts in growing seasons in parts of Africa could leave populations starving, whereas the same change in a wealthy country might have a lesser effect. These shifts will increase demand for food crops globally, further increasing the price of food at home.

Unilateral Planning and Action

DoD, FBI, CIA, DHS: Threats to National Security

Several reports discuss the connection between climate change and security. For example, a report written by 10 retired military leaders, *National Security and the Threat of Climate Change*, concludes that projected climate change represents a serious threat to U.S. national security.[16] International Alert, a nongovernmental organization in the United Kingdom, reports, "The consequences of climate change include a high risk of armed conflict in 46 countries with a total population of 2.7 billion people, and a high risk of political instability in a further 56 countries with a total population of 1.2 billion." Threats to national security will involve a host of agencies, including the DHS, DoD, FBI, CIA, and others.

DoD is affected by disruptions to international peace and stability, which may be exacerbated by the effects of climate change. Figure 7-3 shows that many countries face an increased risk of armed conflict as a secondary or downstream effect of climate change.

The National Intelligence Council completed an assessment of the national security impacts of climate change in June 2008. The details of the report are unavailable (classified), but its publication signals the gravity of this issue.

Climate change-related policies should include consideration of the national security consequences of enacting such policies. Carbon reduction agreements have the potential to add new tensions to already tenuous international relationships.[17]

Figure 7-3. Projected Risk of Armed Conflict as Consequence of Climate Change

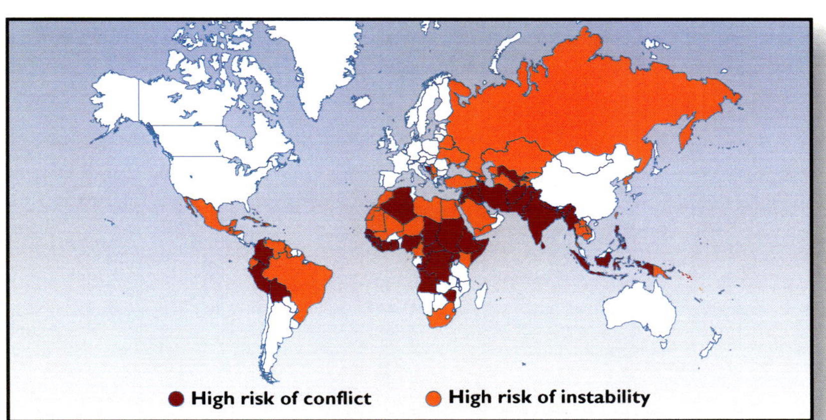

DoD, DOS: Disaster and Evacuation Response for American Citizens Abroad

In addition to providing relief for the citizens of an affected area, the military may assist the DOS with operations to evacuate noncombatants, nonessential military personnel, selected host-nation citizens, and third-country nationals whose lives are in danger from locations in a host foreign nation. A single service or joint task force can carry out noncombatant evacuation operations.[18] The forces deployed to participate in these operations may be needed elsewhere, and these requirements could possibly conflict with other DoD mission needs.

The DOS is likely to have a major role in response to all of the expected international effects of climate change. Not only will overseas ambassadors and foreign service officers have to convey the U.S. policy in each country, they will also have to oversee and participate in the coordination of relief efforts when necessary. Involvement in each country will vary considerably, but simply understanding and conveying the U.S. policy on climate change could mean considerable responsibility for individuals.

DoD and DOS should consider the best available scientific findings on climate change to assist them with short- and long-term planning. Efforts to prepare for these situations should continue.

DoD, USAID: Increased Frequency and Involvement in Disaster Response for Foreign Nations

U.S. agencies are often involved in international disaster response. The U.S. military is often one of the first responders to a major disaster because it can marshal resources quickly and has considerable transport capability. The South Asian Tsunami of 2004, though not attributed to climate change, illustrates the enormity of the efforts made by the United States. Within 48 hours of the tsunami, the U.S. Pacific Command had established a joint task force to coordinate and conduct humanitarian assistance and disaster relief operations. Approximately 16,000 U.S. military personnel deployed for this event. More than two dozen U.S. ships and more than 100 aircraft cooperated in the disaster relief effort at an estimated cost of $5 million a day. The United States spent an estimated $346 million for immediate relief efforts. Another illustrative example of disaster response is Hurricane Mitch, in response to which DoD spent approximately $223 million in FY99, which included immediate life-saving efforts, rehabilitation, and restoration.[19]

If international disasters increase in severity and frequency, our military may be called upon to respond more frequently than it has in the past, putting resource pressure on the organization. Not only will there be increased pressure on human capital and financial resources, but the stakes of successfully addressing disasters may rise as well. The international response to climate change-driven disasters will help determine subsequent societal responses to climate change, such as violent conflict.

U.S. Action with Other Governments

USAID: Assisting with Response to and Preparation for Natural Hazards

The OFDA, in USAID, renders immediate aid and assists communities in preparing for disasters in areas with recurrent natural hazards. The 2006 OFDA annual report shows that the largest portion of the organization's funding went to a single country, Sudan, to cover disruptions including famine and disease. Much of the rest went to Africa and Asia (Figure 7-4).[20] That year, the office responded to 74 disasters, and flooding was the most frequent. This organization may face increased disasters due to climate change, as the numbers of people affected by floods increases dramatically. OFDA, like all government offices, is working with a fixed amount of resources, and the impact of more frequent flooding and other disasters may affect the entire organization and all of its programs.

Figure 7-4. Office of Foreign Disaster Assistance Disaster Funding FY06 by Region

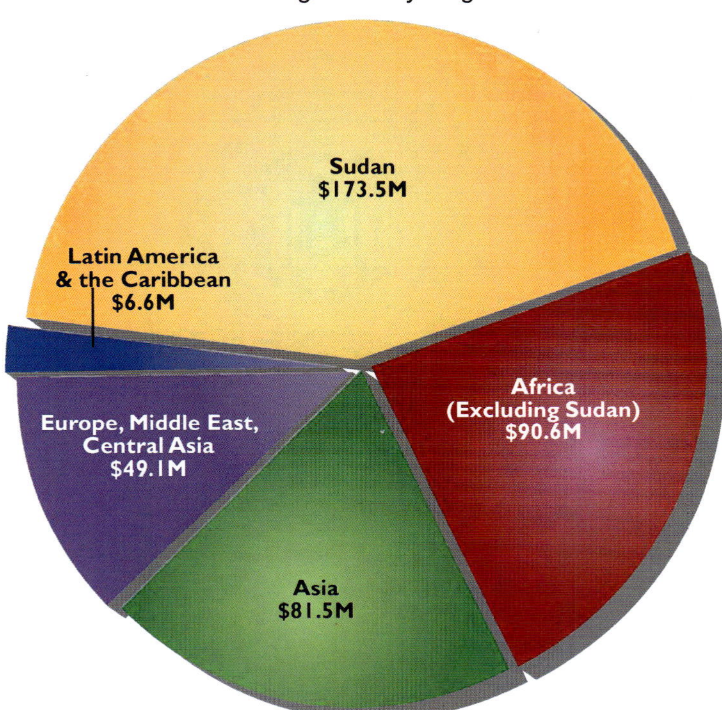

USDA, FTC, DOE: Handling Impacts to Foreign Trade and Production

Downstream effects of climate change may impact domestic organizations. For example, a changing ecosystem in many European regions may shift growing seasons, or change them altogether. Our international trade in food products or energy could be affected, indirectly involving organizations like USDA, the Federal Trade Commission (FTC), and DOE. FTC works with more than 100 foreign authorities charged with regulating competition and consumer protection around the world to support sound policy approaches to doing business abroad.[21] Changes in production and consumption overseas may affect these foreign authorities and, therefore, become relevant for the FTC.

All Agencies: Supply Chain Disruptions

The disruption in foreign trade and production can also have more immediate impacts on federal agencies. Many agencies rely on foreign sources materials, including electronics, textiles, and other materials, to carry out their missions. Climate change-related disruptions could change the ability to acquire these supplies from current sources. Federal managers need to monitor these sources of supply to predict possible disruptions in the supply chain and work with current and alternate suppliers to ensure the agency has the materials necessary to continue operations.

EPA, NASA, NOAA, USGS: Participation or Support to International Research and Discussion

NOAA is the primary funding agency of the International Research Institute (IRI) for Climate and Society at Columbia University. The IRI's mission is to enhance society's capability to understand, anticipate, and manage the impacts of seasonal climate fluctuations to improve human welfare and the environment, especially in developing countries. The IRI has found that in practice, climate risk management strategies in the developing world require dealing with societal and economic issues that span months, seasons, years, and decades.

The need for more research and leadership in coping with climate on all time scales is large and rapidly growing. Much of the developing world will be looking to the United States (and its agencies) to provide this leadership. Our ability to help the developing world cope with climate stress will contribute toward addressing some of the security, disaster response, and other concerns raised in this chapter.

International agreements to which the United States is party may involve a variety of agencies. Any commitment to reduce fossil fuel consumption or any increase in alternative energy will likely affect DOE. International climate change agreements could affect many agencies that work in the environmental sector by requiring them to increase information sharing, scientific research, or R&D. EPA, NOAA, NASA, and USGS may all be called upon to participate in U.S. efforts to comply with international agreements to reduce GHG emissions. Finally, at some point, all federal agencies may be required to reduce GHG emissions, due in some part to U.S. international commitments to do so.

> It is very likely that climate change can slow the pace of progress toward sustainable development either directly through increased exposure to adverse impacts or indirectly through erosion of the capacity to adapt. Over the next half-century, climate change could impede achievement of the Millennium Development Goals.[3]

Impacts from climate change may impede achievement of the UN Millennium Development Goals, a measurement of sustainability agreed on by all countries in an effort to meet the needs of the world's poorest people. This could affect the United States because we have made a (nonbinding) commitment to assist with reaching these goals by 2015, and any obstacles to achieving them could result in the need for more resources to expedite success.

> There is high agreement and much evidence that: Making development more sustainable by changing development paths can make a major contribution to climate change mitigation, but implementation may require resources to overcome multiple barriers. There is a growing understanding of the possibilities to choose and implement mitigation options in several sectors to realize synergies and avoid conflicts with other dimensions of sustainable development.[3]

SUMMARY

The IPCC reports, with very high confidence, climate change will have a variety of impacts on countries around the world—specifically those with vulnerable coastal populations, island nations, and many areas of Europe. The United States will not only have a major role in climate change negotiations, but will also be increasingly involved in responding to the effects of climate change on the international community. U.S. agencies involved in humanitarian aid, conflict resolution, national security, international diplomacy, and international environmental issues are likely to be affected.

In addition to adapting to the effects of climate change in our own nation, federal agencies will have to adapt to the effects elsewhere, as they will have downstream implications for our country across a wide array of sectors and locations.

The responsibilities of federal leaders will also be affected by any agreement resulting from the negotiations at COP 15. Although we cannot predict the components of the agreement, we expect it will increase U.S. government responsibility for reducing emissions in coming years. Climate change and its associated consequences present an unprecedented global challenge. The issue is now of major importance on the world's stage.

---

[1] UN Framework Convention on Climate Change, "The Intergovernmental Panel on Climate Change," *Feeling the Heat*, unfccc.int/essential_background/feeling_the_heat/items/2906.php.

[2] EPA, "International Cooperation," *Climate Change—U.S. Climate Policy*, www.epa.gov/climatechange/policy/internationalcooperation.html.

[3] IPCC key finding. Throughout this book, each purple box shows a key finding from the IPCC with a high level of scientific understanding. See Appendix C for specific AR4 chapter sources for each finding. Chapter 1 gives a full discussion of our selection of key findings.

[4] U.S. DOS, *Bilateral and Regional Climate Partnerships*, www.state.gov/g/oes/climate/c22820.htm.

[5] Harlan Watson, "U.S. Climate Change Policy," (remarks, Seminar of Government Experts, Bonn, Germany, May 16, 2005), unfccc.int/files/meetings/seminar/application/pdf/sem_pre_usa.pdf.

[6] GLOBE International, *Welcome to the Website of GLOBE International*, globeinternational.org.

[7] Asia-Pacific Partnership on Clean Development and Climate, *Welcome to the Asia-Pacific Partnership on Clean Development and Climate*, asiapacificpartnership.org.

[8] Carbon Sequestration Leadership Forum, *What's New*, cslforum.org.

[9] U.S. DOS, *Major Economies Process on Energy Security and Climate Change*, state.gov/g/oes/climate/mem.

[10] National Science and Technology Council Committee on Environment and Natural Resources, *Scientific Assessment of the Effects of Global Change on the United States*, www.climatescience.gov/Library/scientific-assessment/Scientific-AssessmentFINAL.pdf.

[11] IPCC, "Coastal Systems and Low-Lying Areas" in Climate Change *2007: Impacts, Adaptation, and Vulnerability. Working Group II Contribution to the Fourth Assessment Report of the Intergovernmental Panel on Climate Change*, Chapter 6, Cambridge University Press, Cambridge, United Kingdom, and New York, NY, USA, www.ipcc.ch/pdf/assessment-report/ar4/wg2/ar4-wg2-chapter6.pdf.

[12] DefenseAlert, *New Intelligence Assessment Sizes Up National Security Implications of Climate Change*, Inside Washington Publishers, June 18, 2008. (Quotation is from a security briefing delivered by the National Intelligence Council.)

[13] U.S. Census Bureau, Foreign Trade Division, *Foreign Trade Statistics*, www.census.gov/foreign-trade/statistics/index.html, updated October 1, 2007.

[14] Figure 7-2 shows relative percentage changes in precipitation between 2090 and 2099 and 1980 and 1999, using multimodel averages based on the Special Report on Emission Scenarios A1B scenario. The figure on the left shows changes observed December through February, while the one on the right shows those in June through August. Areas with no color indicate less than 66 percent of models agree in the sign of the change. Stippled (dotted) areas are those where more than 90 percent of models agree in the sign of the change.

[15] U.S. Agency for International Development (USAID), "Bilateral Agreements," *Environment*, November 19, 2008, www.usaid.gov/our_work/environment/climate/policies_prog/joint_statements.html.

[16] CNA Corporation, *National Security and the Threat of Climate Change* (Alexandria, VA: CNA Corporation, 2007), securityandclimate.cna.org/report/.

[17] Joshua Busby, *Climate Change and National Security: An Agenda for Action*, Council on Foreign Relations, CSR No. 32, 2007.

[18] Joint Pub. 3-07.5. *Joint Tactics, Techniques, and Procedures for Noncombatant Evacuation Operations,* 1997.

[19] GAO, National Security and International Affairs Division, *Military Operations: DoD's Disaster Relief Assistance in Response to Hurricane Mitch*, Report B-282370, March 29, 1999, archive.gao.gov/paprpdf2/161906.pdf.

[20] USAID, Office of U.S. Foreign Disaster Assistance, *Annual Report for Fiscal Year 2006*, 2007, www.usaid.gov/our_work/humanitarian_assistance/disaster_assistance/publications/annual_reports/pdf/AR2006.pdf.

[21] Federal Trade Commission, *Welcome to the Office of International Affairs*, www.ftc.gov/oia/index.shtml, April 1, 2008.

Appendix A Bibliography

This bibliography contains sources cited in the text of the chapters ("Sources Cited") as well as those considered in the preparation of this book but not cited ("Sources Considered").

Sources Cited

AASC. *Policy Statement on Climate Variability and Change*. 2002. www.stateclimate.org/publications/files/aascclimatepolicy.pdf.

America's Energy Security Trust Fund Act of 2007. H.R. 3416. Proposed by Congressman John Larson, introduced Augu̧st 3, 2007.

Asia-Pacific Partnership on Clean Development and Climate. *Welcome to the Asia-Pacific Partnership on Clean Development and Climate*. asiapacificpartnership.org.

Blasing, T. J. and K. Smith. *Recent Greenhouse Gas Concentrations*. CDIAC. September 2008. cdiac.ornl.gov/pns/current_ghg.html.

Brinkman, Norman, Michael Wang, Trudy Weber, and Thomas Darlington. *Well-to Wheels Analysis of Advanced Fuel/Vehicle Systems—A North American Study of Energy Use, Greenhouse Gas Emissions, and Criteria Pollutant Emissions*. General Motors Corporation, Argonne National Laboratory, and Air Improvement Resource, Inc. May 2005. www.transportation.anl.gov/pdfs/TA/339.pdf.

Busby, Joshua. *Climate Change and National Security: An Agenda for Action*. Council on Foreign Relations, CSR No. 32. 2007.

Business Dictionary. *Sensitivity analysis*. WebFinance Inc. www.businessdictionary.com/definition/sensitivity-analysis.html.

Canes, Michael. *A Comparison of a Cap & Trade System v. Alternative Policies to Curb U.S. Greenhouse Gases*. Washington Roundtable on Science and Public Policy, George C. Marshall Institute. February 22, 2007. p. 12. www.marshall.org/article.php?id=470.

Capehart, Tom. *Cellulosic Biofuels: Analysis of Policy Issues for Congress*. Congressional Research Service (CRS) Report for Congress RL34738. November 7, 2008. fpc.state.gov/documents/organization/112413.pdf.

Carbon Mitigation Initiative, Princeton University. *Stabilization Wedges*. www.princeton.edu/wedges/presentation_resources/.

Carbon Sequestration Leadership Forum. *What's New*. cslforum.org.

Cazenave, Anny. "How Fast Are the Ice Sheets Melting?" *Science*, 314 (5803) November 24, 2006, pp. 1250–1252.

CCSP and the Subcommittee on Global Change Research. *The First State of the Carbon Cycle Report (SOCCR): The North American Carbon Budget and Implications for the Global Carbon Cycle*, Synthesis and Assessment Product

2.2, [King, A.W. et al. (eds.)]. NOAA, National Climatic Data Center, Asheville, NC. www.climatescience.gov/Library/sap/sap2-2/final-report/default.htm.

CCSP. 2007. *Scenarios of Greenhouse Gas Emissions and Atmospheric Concentrations (Part A) and Review of Integrated Scenario Development and Application (Part B).* A Report by the U.S. Climate Change Science Program and the Subcommittee on Global Change Research [Clarke, L., J. Edmonds, J. Jacoby, H. Pitcher, J. Reilly, R. Richels, E. Parson, V. Burkett, K. Fisher-Vanden, D. Keith, L. Mearns, C. Rosenzweig, M. Webster (Authors)]. Department of Energy, Office of Biological & Environmental Research, Washington, DC, USA, 260 pp.

CCSP. 2008. *Analyses of the effects of global change on human health and welfare and human systems.* A Report by the U.S. Climate Change Science Program and the Subcommittee on Global Change Research, Glossary and Acronyms. [Gamble, J.L. (ed.), K.L. Ebi, F.G. Sussman, T.J. Wilbanks, (Authors)]. U.S. Environmental Protection Agency, Washington, DC, USA. www.climatescience.gov/Library/sap/sap4-6/final-report/.

CCSP. 2008. *Impacts of Climate Change and Variability on Transportation Systems and Infrastructure: Gulf Coast Study.* Phase I, Synthesis and Assessment Product 4.7. [Savonis, M. J., V.R. Burkett, and J.R. Potter (eds.)]. March 2008, 445 pp. www.climatescience.gov/Library/sap/sap4-7/final-report/.

CCSP. 2008. *The effects of climate change on agriculture, land resources, water resources, and biodiversity in the United States.* A Report by the U.S. Climate Change Science Program and the Subcommittee on Global Change Research, by P. Backlund et al. U.S. Department of Agriculture, Washington, DC., USA, 362 pp. www.climatescience.gov/Library/sap/sap4-3/final-report/Synthesis_SAP_4.3.pdf.

CCSP. 2008. *Our Changing Planet, the US Climate Change Science Program (CCSP) for Fiscal Year 2009.* A Report by the U.S. Climate Change Science Program and the Subcommittee on Global Change Research, A Supplement to the President's Budget for FY 2009. www.usgcrp.gov/usgcrp/Library/ocp2009/default.htm.

CCSP. 2008. *Preliminary review of adaptation options for climate-sensitive ecosystems and resources*, A Report by the U.S. Climate Change Science Program and the Subcommittee on Global Change Research. S. H. Julius and J.M. West, eds. (Washington, DC: USEPA, 2008), 873 pp. www.climatescience.gov/Library/sap/sap4-4/final-report/.

CCSP. 2008. *Weather and Climate Extremes in a Changing Climate. Regions of Focus: North America, Hawaii, Caribbean, and U.S. Pacific Islands.* A Report by the U.S. Climate Change Science Program and the Subcommittee on Global Change Research. [Thomas R. Karl, Gerald A. Meehl, Christopher D. Miller, Susan J. Hassol, Anne M. Waple, and William L. Murray (eds.)]. Department of Commerce, NOAA's National Climatic Data Center, Washington, DC, USA, 164 pp., www.climatescience.gov/Library/sap/sap3-3/final-report/default.htm.

CDC, HHS. *Cost Effectiveness Analysis*. www.cdc.gov/owcd/EET/CostEffect2/Fixed/Glossary.html#sensitivityanalysis.

Ceres and the Investor Network on Climate Risk. *Managing the Risks and Opportunities of Climate Change: A Practical Toolkit for Corporate Leaders*. 2006. Boston, January 2006. www.pewclimate.org/docUploads/Ceres_corportate_toolkit_012006_010507_130824.pdf.

Climate Change Technology Program, www.climatetechnology.gov/.

CNA Corporation. *National Security and the Threat of Climate Change* (Alexandria, VA: CNA Corporation, 2007). securityandclimate.cna.org/report/.

Coastal Zone Management Subgroup, IPCC Response Strategies Working Group. *Strategies for Adaption to Sea Level Rise*. November 1990. www.epa.gov/climatechange/effects/coastal/SLRAdaption.html.

Cooper, Richard N. *Alternatives to Kyoto: the Case for a Carbon Tax*. Working Paper, Weatherhead Center for International Affairs, Harvard University. 2006. www.economics.harvard.edu/faculty/cooper/files/Kyoto_ct.pdf.

Creyts, J., A. Derkach, S. Nyquist, K. Ostrowski, and J. Stephenson. *Reducing U.S. Greenhouse Gas Emissions: How Much at What Cost?* Executive Report of the U.S. Greenhouse Gas Abatement Mapping Initiative, McKinsey & Company. December 2007. www.mckinsey.com/clientservice/ccsi/greenhousegas.asp.

Crutzen, P.J., et al. "N_2O release from agro-biofuel production negates global warming reduction by replacing fossil fuels." *Atmospheric Chemistry and Physics*, Vol. 8, No. 2. January 2008.

DefenseAlert. *New Intelligence Assessment Sizes Up National Security Implications of Climate Change*. Inside Washington Publishers. June 18, 2008. (Quotation is from a security briefing delivered by the National Intelligence Council.)

Doan, Lurita. (statement by Administrator, GSA, before the Committee on Transportation and Infrastructure, U.S. House of Representatives, May 11, 2007). www.gsa.gov/Portal/gsa/ep/contentView.do?pf=y&noc=T&channelId=-18801&contentId=23117&programId=9641&pageTypeId=8199&contentType=GSA_BASIC.

DOC, NOAA, and the National Ocean Service (NOS). *Population Trends Along the United States Coast, 1980-2008*. Coastal Trends Report Series. 2005. oceanservice.noaa.gov/programs/mb/supp_cstl_population.html.

DOE and EPA. *ENERGY STAR*. www.energystar.gov/.

DOE Consortium. *Carbon Sequestration in Terrestrial Ecosystems*. csite.esd.ornl.gov/.

DOE, E.O. Lawrence Berkeley National Laboratory, University of California, Environmental Energy Technologies Division (EETD). "Facts about U.S. Paper Use." *Cutting Paper*. eetd.lbl.gov/paper/ideas/html/uspaperuse.htm.

DOE, EIA. *Annual Energy Review 2007*. DOE/EIA-0384(2007), June 23, 2008. www.eia.doe.gov/emeu/aer/.

DOE, EIA. *Biofuels in the U.S. Transportation Sector*. February 2007. www.eia.doe.gov/oiaf/analysispaper/biomass.html.

DOE, EIA. *Emissions of Greenhouse Gases in the United States*. DOE/EIA-0573, 2006 (Washington, DC: DOE, November 28, 2007). www.eia.doe.gov/oiaf/1605/ggrpt/.

DOE, EIA. *International Energy Outlook 2008, Highlights*. www.eia.doe.gov/oiaf/ieo/pdf/highlights.pdf.

DOE, FEMP, *You Have the Power Campaign Materials*, August 15, 2008, www1.eere.energy.gov/femp/services/yhtp/campaign_materials.html.

DOE. "Energy Efficient Products." *Federal Energy Management Program*. www1.eere.energy.gov/femp/procurement/eep_requirements.html.

DOE. "EnergyPlus Energy Simulation Software." *Energy Efficiency and Renewable Energy*. November 15, 2008. www.eere.energy.gov/buildings/energyplus/.

DOE. "Federal Energy Management Program." *Energy Efficiency and Renewable Energy*. www1.eere.energy.gov/femp/index.html.

DOE. "You Have the Power." *Federal Energy Management Program*. April 19, 2007. www1.eere.energy.gov/femp/services/yhtp/.

DOL, Bureau of Labor Statistics. "Federal Government, Excluding the Postal Service." *Career Guide to Industries, 2008-09 Edition*. www.bls.gov/oco/cg/cgs041.htm.

Emanuel, Kerry. "Increasing destructiveness of tropical cyclones over the past 30 years." *Nature*, Vol. 436, No. 7051 (August 4, 2005), pp. 686–688, ISSN 0028-0836.

Environmental Administration of Finland, *Environmentally related energy taxation in Finland*, October 10, 2008, www.environment.fi/default.asp?contentid=147208&lan=en.

EO 13423. *Strengthening Federal Environmental, Energy, and Transportation Management*. President George W. Bush, January 24, 2007. www.whitehouse.gov/news/releases/2007/01/20070124-2.html.

EO 13432. *Cooperation Among Agencies in Protecting the Environment with Respect to Greenhouse Gas Emissions From Motor Vehicles, Nonroad Vehicles, and Nonroad Engines*. President George W. Bush, May 14, 2007. www.whitehouse.gov/news/releases/2007/05/20070514-1.html.

EPA. *Climate Change*, www.epa.gov/climatechange/index.html.EPA. "International Cooperation." *Climate Change—U.S. Climate Policy*. www.epa.gov/climatechange/policy/internationalcooperation.html.

EPA. *Economics of Biofuels*. yosemite.epa.gov/EE/epa/eed.nsf/pages/Biofuels.html.

EPA. *Glossary of Climate Change Terms*. www.epa.gov/climatechange/glossary.html.

EPA. "About WasteWise." *Wastes—Partnerships—WasteWise Program*. October 7, 2008. wastewise.tms.icfi.com/wisesearch/results.asp.

EPA. "Coastal Zones and Sea Level Rise." *Climate Change—Health and Environmental Effects*. August 19, 2008. www.epa.gov/climatechange/effects/coastal/index.html.

EPA. "ENERGY STAR For Commercial Real Estate." *ENERGY STAR*. www.energystar.gov/index.cfm?c=comm_real_estate.bus_comm_realestate.

EPA. "Temperature Changes." *Climate Change—Science*. March 31, 2008. www.epa.gov/climatechange/science/recenttc.html.

EPA. "The ENERGY STAR Challenge." *ENERGY STAR*. www.energystar.gov/index.cfm?c=challenge.bus_challenge.

EPA. "U.S. Climate Policy and Actions." *Climate Change—U.S. Climate Policy*. December 20, 2007. www.epa.gov/climatechange/policy/index.html.

EPA. *Climate Change Images*. www.epa.gov/climatechange/science/images/co2-temp.gif.

EPA. *Climate Leaders*. November 19, 2008. www.epa.gov/stateply/.

EPA. *Future Sea Level Changes*. December 20, 2007. www.epa.gov/climatechange/science/ futureslc.html.

EPA, *Glossary of Climate Change Terms*, www.epa.gov/climatechange/glossary.html.

EPA. *Inventory of U.S. Greenhouse Gas Emissions and Sinks: 1990–2006*. USEPA #430-R-08-005. April 2008. www.epa.gov/climatechange/emissions/usinventoryreport.html.

EPA. *Milestones in Garbage: A historical timeline of municipal solid waste management*. www.epa.gov/epaoswer/non-hw/muncpl/timeline_alt.htm.

EPA. Policy, Planning, and Evaluation and Maine State Planning Office. *Anticipatory Planning For Sea-Level Rise Along The Coast of Maine*. EPA-230-R-95-900. September 1995. www.epa.gov/climatechange/effects/downloads/maine_0.pdf.

EPA. *SmartWay*. October 30, 2008. www.epa.gov/smartway/index.htm.

EPA. *The Science of Ozone Layer Depletion*. www.epa.gov/ozone/science/index.html. October 4, 2007.

European Climate Exchange. *Historical data—ECX EUA Futures Contract*. 2008. www.europeanclimateexchange.com/default_flash.asp.

Federal Electronics Challenge. www.federalelectronicschallenge.net/index.htm.

Federal Jobs Network. federaljobs.net/about.htm.

Federal Network for Sustainability. *Greening Federal Copier Paper Wizard*. www.federalsustainability.org/initiatives/gfcp.htm.

Federal Network for Sustainability. *USEPA Region 10—Paper Reduction Case Study*. December 2006. www.federalsustainability.org/initiatives/gfcp/EPA10papercasestudy.pdf.

FTC. *Welcome to the Office of International Affairs*. April 1, 2008. www.ftc.gov/oia/index.shtml.

GAO, National Security and International Affairs Division. *Military Operations: DoD's Disaster Relief Assistance in Response to Hurricane Mitch*. Report B-282370, March 29, 1999. archive.gao.gov/paprpdf2/161906.pdf.

GAO. *Air Pollution: Allowance Trading Offers an Opportunity to Reduce Emissions at Less Cost*. GAO/RCED-95-30. Report to the Chairman, Environment, Energy, and Natural Resources Subcommittee, Committee on Government Operations, House of Representatives (Washington, DC, December 1994), p. 37. www.gao.gov/archive/1995/rc95030.pdf.

GAO. *Carbon Offsets: The U.S. Voluntary Market Is Growing, but Quality Assurance Poses Challenges for Market Participants*. GAO-08-1048. September 26, 2008. www.gao.gov/htext/d081048.html.

GAO. *Climate Change: Financial Risks to Federal and Private Insurers in Coming Decades Are Potentially Significant*. Report to the Committee on Homeland Security and Governmental Affairs, U.S. Senate, by John B. Stephenson et al. GAO-07-285. March 2007. www.gao.gov/new.items/d07285.pdf.

GAO. *Electronic Waste: Strengthening the Role of the Federal Government in Encouraging Recycling and Reuse*. GAO-06-47. November 2005. www.gao.gov/new.items/d0647.pdf.

GLOBE International. *Welcome to the website of GLOBE International*. globeinternational.org.

GSA. *Federal Fleet Report: Fiscal Year 2007*. January 31, 2008. www.gsa.gov/graphics/ogp/FFR2007_508.pdf.

GSA. *U.S. General Services Administration*. www.gsa.gov.

Hamlet, A.F., P.W. Mote, M.P. Clark, D.P. Lettenmaier. "20th Century Trends in Runoff, Evapotranspiration, and Soil Moisture in the Western U.S." *Journal of Climate*, Vol. 20, No. 8, April 2007, pp. 1468–1486.

HM Revenue and Customs. *Climate Change Levy (CCL): Changes to rates*. BN HMRC68/07. www.hmrc.gov.uk/thelibrary/rates.htm.

ICLEI—Local Governments for Sustainability. 2008. www.iclei.org/index.php?id=iclei-home&no_cache=1.

IPCC. "Fact Sheet: How the Summary for Policymakers of an IPCC Working Group report is approved during the Plenary." *Information for the press*. www.ipcc.ch/press/ar4-factsheet2.htm.

IPCC. "Procedures for the Preparation, Review, Acceptance, Adoption, Approval, and Publication of IPCC Reports." *Principles Governing IPCC Work*, Appendix A. Adopted at the Fifteenth Session (San Jose, April 15- 18, 1999), amended at the Twentieth Session (Paris, February19-21, 2003) and Twenty-first Session (Vienna, November 3 and 6-7, 2003). www.ipcc.ch/pdf/ipcc-principles/ipcc-principles-appendix-a.pdf.

IPCC. *Climate Change 2001: Synthesis Report. A Contribution of Working Groups I, II and III to the Third Assessment Report of the Intergovernmental Panel on Climate Change*. Cambridge University Press, Cambridge, United Kingdom, and New York, NY, USA. www.ipcc.ch/ipccreports/tar/vol4/english/index.htm.

IPCC. *Climate Change 2007: Impacts, Adaptation and Vulnerability. Working Group II Contribution to the Fourth Assessment Report of the Intergovernmental Panel on Climate Change*. Cambridge University Press, Cambridge, United Kingdom, and New York, NY, USA. www.ipcc.ch/ipccreports/ar4-wg2.htm.

IPCC. *Climate Change 2007: Mitigation of Climate Change. Working Group III Contribution to the Fourth Assessment Report of the Intergovernmental Panel on Climate Change*. Cambridge University Press, Cambridge, United Kingdom, and New York, NY, USA, www.ipcc.ch/ipccreports/ar4-wg3.htm.

IPCC. *Climate Change 2007: Synthesis Report. Contribution of Working Groups I, II and III to the Fourth Assessment Report of the Intergovernmental Panel on Climate Change*. Cambridge University Press, Cambridge, United Kingdom, and New York, NY, USA. www.ipcc.ch/ipccreports/ar4-syr.htm.

IPCC. *Climate Change 2007: The Physical Science Basis. Working Group I Contribution to the Fourth Assessment Report of the Intergovernmental Panel on Climate Change*. Cambridge University Press, Cambridge, United Kingdom, and New York, NY, USA. www.ipcc.ch/ipccreports/ar4-wg1.htm

IPCC. *IPCC Special Report on Carbon Dioxide Capture and Storage*. Prepared by Working Group III of the Intergovernmental Panel on Climate Change [Metz, B., O. Davidson, H. C. de Coninck, M. Loos, and L. A. Meyer (eds.)]. Cambridge University Press, Cambridge, United Kingdom and New York, NY, USA, 442 pp. 2005. www.ipcc.ch/pdf/special-reports/srccs/srccs_wholereport.pdf.

IPCC/TEAP. 2005. *Safeguarding the Ozone Layer and the Global Climate System: Issues related to hydrofluorocarbons and perfluorocarbons*. by Bert Metz, et al. (Eds). Cambridge University Press, U.K., pp. 478. www.ipcc.ch/pdf/special-reports/sroc/sroc_full.pdf.

Joint Pub. 3-07.5. *Joint Tactics, Techniques, and Procedures for Noncombatant Evacuation Operations*. 1997.

Lemmen, D. S. and F. J. Warren, eds. *Climate Change Impacts and Adaptation: A Canadian Perspective* (Natural Resources Canada, 2004), 174 pp. adaptation.nrcan.gc.ca/perspective/pdf/report_e.pdf.

Letter from Michael Wang, Center for Transportation Research, Argonne National Laboratory, and Zia Haq, Office of Biomass Program, OEERE, DOE, to *Science*, March 14, 2008. www.transportation.anl.gov/pdfs/letter_to_science_anldoe_03_14_08.pdf.

Lieberman-Warner Climate Security Act of 2007. Senate Bill 2191. Proposed by Senators Joe Lieberman (I-CT) and John Warner (R-VA), introduced October 18, 2007.

M. J. Bradley & Associates. *Comparison of Energy Use and CO_2 Emissions From Different Transportation Modes.* For the American Bus Association, Manchester, NH. May 2007. www.buses.org/files/ComparativeEnergy.pdf.

Marland, G., T.A. Boden, and R.J. Andres. "Global CO_2 Emissions from Fossil Fuel Burning, Cement Manufacture, and Gas Flaring, 1751–2004." In *Trends: A Compendium of Data on Global Change.* CDIAC. cdiac.esd.ornl.gov/ftp/ndp030/region.1751_2004.ems.

Massachusetts Water Resources Authority (MWRA). *The Deer Island Sewage Treatment Plant.* www.mwra.state.ma.us/03sewer/html/sewditp.htm.

Mayor's Office of Operations, Office of Long-term Planning and Sustainability. *Inventory of New York City Greenhouse Gas Emissions.* Mayor of New York. April 2007. home2.nyc.gov/html/planyc2030/downloads/pdf/emissions_inventory.pdf.

Mehdi, B., Charles Mrena, and Al Douglas. *Adapting to Climate Change: An Introduction for Canadian Municipalities.* Canadian Climate Impacts and Adaptation Research Network (C-CIARN). 32 pp. February 2006. www.adaptation.nrcan.gc.ca/municipalities_e.php.

Metroeconomica Limited. *Costing the Impacts of Climate Change in the UK: Implementation Guidelines.* Final Report for The UK Climate Impacts Programme (UKCIP). June 18, 2004. www.ukcip.org.uk/images/stories/Pub_pdfs/CostingsImplementationGuidelines.pdf.

Miller, K. and D. Yates. *Climate Change and Water Resources: a Primer for Municipal Water Providers.* American Water Works Association and University Corporation for Atmospheric Research. Project #2973. 2006. www.awwarf.org/research/topicsandprojects/projectSnapshot.aspx?pn=2973.

NASA. "Gross Domestic Product Deflator Inflation Calculator." *Cost Estimating Web Site.* cost.jsc.nasa.gov/inflateGDP.html.

National Science and Technology Council Committee on Environment and Natural Resources. *Scientific Assessment of the Effects of Global Change on the United States.* www.climatescience.gov/Library/scientific-assessment/Scientific-AssessmentFINAL.pdf.

NOAA Coastal Services Center. "Discounting and Time Preference." *Restoration Economics.* www.csc.noaa.gov/coastal/economics/discounting.htm.

NOAA, National Climate Data Center. *Global Surface Temperature Anomalies.* February 6, 2006. www.ncdc.noaa.gov/oa/climate/research/anomalies/anomalies.html.

NOAA, National Climatic Data Center. *Global Warming Frequently Asked Questions.* www.ncdc.noaa.gov/oa/climate/globalwarming.html.

NOAA, National Climatic Data Center. *Vision Figure 2.* www.climatescience.gov/Library/stratplan2003/vision/VisionFig2.jpg.

NOAA. *State of the Climate in 2007.* www.ncdc.noaa.gov/oa/climate/research/2007/ann/bams/.

NREL. *Energy Lab Sets Aggressive Greenhouse Gas Reduction Goal: NREL pledges to cut carbon footprint, impact on environment by 75 percent.* NREL News Release NR-2907. December 4, 2007. www.nrel.gov/news/press/2007/549.html.

NREL. *Geothermal Energy Basics, Learning about Renewable Energy,* www.nrel.gov/learning/re_geothermal.html.

O'Hare, Michael. "Greenhouse Gas Emissions from Indirect Land Use Change." presentation to the California Air Resources Board (CARB) Low-Carbon Fuel Standard (LCFS) Working Group 3, Sacramento, CA. January 17, 2008. www.arb.ca.gov/fuels/lcfs/011708UCBLUCcolor.pdf.

Ohlson, Dan W., Greg A. McKinnon, and Kelvin G. Hirsch. "A structured decision-making approach to climate change adaptation in the forest sector." *The Forestry Chronicle*, January/February 2005, Vol. 81, No. 1. www.c-ciarn.ca/pdf/forestry_chronicle_article_published.pdf.

OMB, Budget Concepts Branch. *Object Class Analysis: Budget of the U.S. Government, Fiscal Year 2009.* www.whitehouse.gov/omb/budget/fy2009/pdf/objclass.pdf.

Oxford University Press. *New Oxford American Dictionary.* Second Edition. Erin McKean (Ed.), 2051 pp. May 2005. ISBN 0-19-517077-6.

Pacala, Steven, and Robert Socolow. "Stabilization Wedges: Solving the Climate Problem for the Next 50 Years with Current Technologies." *Science*, Vol. 305, 968-972 (August 13, 2004). www.carbonsequestration.us/Papers-presentations/htm/Pacala-Socolow-ScienceMag-Aug2004.pdf.

Perez, Rosa T. et al. "Continuing the Adaptation Process." in *Policy Frameworks for Climate Change: Developing Strategies, Policies and Measures.* United Nations Development Program and Global Environmental Facility. www.undp.org/gef/undp-gef_publications/publications/apf%20technical%20paper09.pdf.

Pew Center on Global Climate Change and the Pew Center on the States. "Climate Change 101: Adaptation" part of the *Climate Change 101: Understanding and Responding to Global Climate Change* series. www.pewclimate.org/docUploads/Adaptation_0.pdf.

Pizer, William A., Richard D. Morgenstern, and Jhih-Shyang Shih. *Evaluating Voluntary Climate Programs in the United States.* Resources for the Future. Discussion Paper 08-13. July 2008. www.rff.org/Publications/Pages/PublicationDetails.aspx?PublicationID=20485.

Pyke, Christopher R., Raymond Najjar, et al. *Climate Change and the Chesapeake Bay: State-of-the-Science Review and Recommendations.* A Report from the Chesapeake Bay Program Science and Technical Advisory Committee (STAC). April 21, 2008.

www.chesapeake.org/stac/QuarterlyMeetingPresentationsMaterial/june08/climatechangefinaldraft.pdf.

Raffensperger, Lisa. *A Fresh Green for No-Till Farming*. World Resources Institute. 2008. earthtrends.wri.org/updates/node/286.

Ramseur, Jonathan L. *The Role of Offsets in a Greenhouse Gas Emissions Cap-and-Trade Program: Potential Benefits and Concerts*. Congressional Research Service, CRS Report for Congress, RL34436. April 4, 2008. fpc.state.gov/documents/organization/103675.pdf.

Reed Business Information. "White Paper on Sustainability." a supplement to *Building Design & Construction*. p. 20–21. www.bdcnetwork.com/article/CA6566572.html?nid=2886.

Regional Greenhouse Gas Initiative. www.rggi.org.

Save Our Climate Act of 2007. H.R. 2069. Proposed by Congressman Pete Stark of California, introduced April 26, 2007.

Searchinger, Timothy D., et al. "Use of U.S. Croplands for Biofuels Increases Greenhouse Gases Through Emissions from Land-Use Change." *Science*, February 29, 2008. Vol. 319.

Smith, Dan and Janani Vivekanda. "A Climate of Conflict: The Links Between Climate Change, Peace and War." *International Alert*, November 2007. www.international-alert.org/pdf/A_Climate_Of_Conflict.pdf.

Socolow, R. H. *Introduction to the Wedges*. Carbon Mitigation Initiative, Princeton Environmental Institute. www.princeton.edu/~cmi/resources/stabwedge.htm.

Steffen, Konrad and Russell Huff. *Greenland Melt Extent*. Cires Steffan Research Group, Boulder, CO. September 28, 2005. cires.colorado.edu/science/groups/steffen/greenland/melt2005.

Stephenson, John B. et al. *Climate Change: Agencies Should Develop Guidance for Addressing the Effects on Federal Land and Water Resources*. GAO-07-863 (Washington, DC: GAO, August 2007). www.gao.gov/new.items/d07863.pdf.

Stern, Nicholas. *The Economics of Climate Change: The Stern Review*. October 30, 2006, 700 pp. www.hm-treasury.gov.uk/sternreview_index.htm.

The Carbon Trust. *Local authorities—Saving energy in local authority buildings*. Sector Overview CTV028. London, UK. June 2007. www.carbontrust.co.uk/carbon/publicsector/la/.

The Climate Registry. www.theclimateregistry.org.

The Greenhouse Gas Protocol Initiative. 2008. www.ghgprotocol.org.

The NASA Earth Observatory. *The Carbon Cycle: The Human Role*. 2001. earthobservatory.nasa.gov/Library/CarbonCycle/carbon_cycle4.html.

The NASA Earth Observatory. *The Earth's Energy Balance*. earthobservatory.nasa.gov/Experiments/PlanetEarthScience/GlobalWarming/GW_Movie2.php.

The United Nations Environmental Programme (UNEP)/GRID-Arendal. 2005. *Vital Climate Change Graphics*. UNEP/GRID-Arendal, Arendal, Norway, Nairobi, Kenya, ISBN 82-7701-031-1, p. 6. www.grida.no/_res/site/file/publications/vital-climate_change_update.pdf.

Titius, James G., Richard A. Park, Stephen P. Leatherman, J. Richard Weggel, Michael S. Greene, Paul W. Mausel, Scott Brown, Gary Gaunt, Manjit Trehan, and Gary Yohe. "Greenhouse Effect and Sea Level Rise: The Cost of Holding Back the Sea." *Coastal Management* (1991), Volume 19, 171-204. www.epa.gov/climatechange/effects/coastal/slrmaps_cost_of_holding.html#abstract.

U.S. Census Bureau, Foreign Trade Division. *Foreign Trade Statistics*. www.census.gov/foreign-trade/statistics/index.html. October 1, 2007.

U.S. Census Bureau. *Annual Projections of the Total Resident Population as of July 1: Middle, Lowest, Highest, and Zero International Migration Series, 1999 to 2100, NP-T1*. January 2000. www.census.gov/population/projections/nation/summary/np-t1.pdf.

U.S. Climate Change Technology Program. *Strategic Plan*. DOE/PI-0005. September 2006.

U.S. Department of State, Bureau of Population, Refugees and Migration (PRM). *Writing Goals, Objectives, and Indicators for PRM-funded Humanitarian Assistance Projects*. www.state.gov/g/prm/109569.htm.

U.S. Fairchild Air Force Base. *Statement of Work for Base Custodial Service*. April 19, 2005. www.ofee.gov/gp/custodialsow.pdf.

UNFCCC. "The Intergovernmental Panel on Climate Change." *Feeling the Heat*. unfccc.int/essential_background/feeling_the_heat/items/2906.php.

US Department of State. *Bilateral and Regional Climate Partnerships*. www.state.gov/g/oes/climate/c22820.htm.

US Department of State. *Major Economies Process on Energy Security and Climate Change*. state.gov/g/oes/climate/mem.

USAID, Office of U.S. Foreign Disaster Assistance. *Annual Report for Fiscal Year 2006*. 2007. www.usaid.gov/our_work/humanitarian_assistance/disaster_assistance/publications/annual_reports/pdf/AR2006.pdf.

USAID. "Bilateral Agreements." *Environment*, November 19, 2008. www.usaid.gov/our_work/environment/climate/policies_prog/joint_statements.html.

USGCRP, National Assessment Synthesis Team. "Overview: Coastal Areas and Marine Resources." *Climate Change Impacts on the United States: The Potential Consequences of Climate Variability and Change*. 2000. www.usgcrp.gov/usgcrp/Library/nationalassessment/overviewcoastal.htm.

USGCRP. *About the US Global Change Research Program*. www.usgcrp.gov/usgcrp/about/default.htm. June 22, 2007.

USGCRP. *Announcements from the U.S. Climate Change Science Program.* November 12, 2008. www.usgcrp.gov/usgcrp/default.php.

USGCRP. *Strategic Plan for the U.S. Climate Change Program.* July 2003 (update May 2008).

USGS. *The Water Cycle: Transpiration.* ga.water.usgs.gov/edu/watercycletranspiration.html.

Watson, Harlan. "US Climate Change Policy." (remarks, Seminar of Government Experts, Bonn, Germany, May 16, 2005). unfccc.int/files/meetings/seminar/application/pdf/sem_pre_usa.pdf.

World Resources Institute and World Business Council for Sustainable Development. *Greenhouse Gas Protocol, A Corporate Accounting and Reporting Standard, Revised Edition.* Geneva, Switzerland, and Washington DC. www.ghgprotocol.org/files/ghg-protocol-revised.pdf.

World Resources Institute. *Comparison of Legislative Climate Change Targets.* John Larsen and Robert Heilmayr. September 9, 2008.

Sources Considered

"Filling it All Back Up Again: Carbon Capture and Storage." *The Economist.* March 8, 2008: 65.

Alcorn, J.M, Rosenzweig, C.E., Horton, R.M. and Higuchi, I.S. *U.S. Space Transportation and Climate Change: Potential Impacts of Climate Change on Access to Space.* A Report from the 11th ASCE Aerospace Division International Conference (Earth and Space 2008), Long Beach, CA. March 3-6, 2008.

Asia-Pacific Partnership on Clean Development and Climate. *E-Newsletter.* Vol 1, Nr 1., March 2008. www.asiapacificpartnership.org.

Australian Government, Department of the Environment, Water, Heritage, and the Arts. *Staff Tools - A Big Plus for Morale.* March 4, 2008. www.environment.gov.au/settlements/challenge/members/stafftools.html.

Brown, Marilyn A., Frank Southworth, Andrea Sarzynski. "Shrinking the Carbon Footprint of Metropolitan America." *Blueprint for American Prosperity Series.* Metropolitan Policy Program, Brookings Institute. May 2008. www.blueprintprosperity.org.

California Energy Commission. *Climate Change Impacts and Adaptation in California.* In Support of the 2005 Integrated Energy Policy Report Proceeding (Docket # 04-IEPR-01E), Guido Franco, et al. June 2005. www.ccap.org/docs/resources/427/Climate_Change_Impacts_and_Adaptation_in_California.pdf.

Carbon Sequestration Leadership Forum (CSLF). *A Carbon Capture and Storage Technology Initiative Charter.*

CCSP. 2006. *Preliminary Review of Adaptation Options for Climate-Sensitive Ecosystems and Resources.* Final Prospectus for CCSP Synthesis and Assessment

Product 4.4. EPA, Washington, DC. July 31, 2006. www.climatescience.gov/Library/sap/sap4-4/sap4-4prospectus-final.htm.

CCSP. 2006. *U.S. Climate Change Report*. Chapter 6, Vulnerability Assessment, Climate Change Impacts and Adaptation Measures, Washington, DC. pp 67-75.

CCSP. 2008. *Scientific Assessment of the Effects of Global Change on the United States*. A Report of the Committee on Environment and Natural Resources. National Science and Technology Center, Washington, DC. May 2008. 261 pp. www.climatescience.gov/Library/scientific-assessment/Scientific-AssessmentFINAL.pdf.

Commission of the European Communities. *Adapting to Climate Change in Europe - Options for EU Action*. Green Paper from the Commission to the Council. The European Parliament, the European Economic and Social Committee and the Committee of the Regions. Brussels. June 2005.

Congressional Budget Office. *Policy Options for Reducing CO_2 Emissions*. The Congress of the United States, Pub. No. 2930. February 2008. www.cbo.gov/ftpdocs/89xx/doc8934/02-12-Carbon.pdf

Connaughton, J. and P. Dobriansky. *Press Briefing on Major Economies Meeting*. (Chairman, White House Council on Environmental Quality, and Under Secretary of State for Democracy and Global Affairs, April 18, 2008). www.whitehouse.gov/news/releases/2008/04/20080418-20.html.

Cossa, Ralph A. "South Asian Tsunami: US Military Provides 'Logistical Backbone' For Relief Operation." *eJournalUSA*. March 2005.

Cossa, Ralph A. "Update Report: South Asian Tsunami: U.S. Military Provides 'Logistical Backbone' for Relief Operation." *Foreign Policy Agenda*. November 2007.

Council of Australian Governments. *National Climate Change Adaptation Framework*. February 2006. www.coag.gov.au/coag_meeting_outcomes/2007-04-13/docs/national_climate_change_adaption_framework.pdf.

Del Pino, Putt, Samantha, Ryan Levinson, John Larsen. *Hot Climate, Cool Commerce: A Service Sector Guide to Greenhouse Gas Management*. A Report of the World Resources Institute, ed. Margaret B. Yamashita, Washington, DC. May 2006.

DOE, EIA. *Glossary*. www.eia.doe.gov/glossary/index.html.

DOI, NationalAtlas.gov. *Federal Lands and Indian Reservations*. April 29, 2008 nationalatlas.gov/printable/fedlands.html.

DOI. *Adaptive Management Technical Guide: Chapter 2, When Should Adaptive Management be Used?*. Washington, DC, pp 9-19. www.doi.gov/initiatives/AdaptiveManagement/TechGuide/Chapter2.pdf.

Drexhage, John, Deborah Murphy, et al. *Climate Change and Foreign Policy: An Exploration of Options for Greater Integration*. International Institute for Sustainable Development. Winnipeg: Unigraphics Ltd, 2006. www.iisd.org/publications/pub.aspx?id=843.

Energy Policy Act. H.R. 776, (1992), Public Law No: 102-486. thomas.loc.gov/cgi-bin/query/z?c102:H.R.776.ENR:.

EPA. "Energy." *Inventory of U.S. Greenhouse Gas Emissions and Sinks: 1998-2005*. Washington, DC. 2007.

EPA. *Energy Star Partnership*. www.energystar.gov/index.cfm?fuseaction=globalwarming.showPledgeHome&topOrgCategory=GOV.

EPA. *U.S. Emissions Inventory*. 2005.

EPA. *U.S. Emissions Inventory*. 2006.

EPA. *WasteWise 2005 Annual Report: Forging Ahead*. December 2005. Washington, DC.

European Environment Agency. *Vulnerability and Adaptation to Climate Change in Europe*. EEA Technical Report ISSN 1725-2237, Copenhagen. July 2005. reports.eea.europa.eu/technical_report_2005_1207_144937/en/EEA_Technical_report_7_2005.pdf.

GAO. *Climate Change: Expert Opinion on the Economics of Policy Options to Address Climate Change*. Report to Congressional Requestors, GAO-08-605. Washington, DC. May 2008.

GAO. *Expert Opinion on the Economics of Policy Options to Address Climate Change*. GAO 08-605. May 2008.

Ginocchio, Alaine, Kevin L. Doran et al. *The Boundaries of Executive Authority: Using Executive Orders to Implement Federal Climate Change Policy*. A Report of the Center for Energy & Environmental Security, University of Colorado Law School. February 2008.

Hamilton, James D. Home Page. Professor of Economics, University of California, San Diego. www.econ.ucsd.edu/~jhamilto/.

Hartmann, Betsy. "War Talk and Climate Change." *truthout.org*. November 26, 2007. www.truthout.org/article/betsy-hartmann-war-talk-and-climate-change.

Huntington, Henry, G. Weller, E. Bush, T. Callaghan, V. Kattsov, M. Nuttall. "An Introduction to the Arctic Climate Impact Assessment" In *Arctic Climate Impact Assessment*. ACIA. pp. 1-20. Cambridge University Press, New York, NY, USA, ISBN 0-521-86509-3, 2005. www.acia.uaf.edu/pages/scientific.html.

International Charter Space and Major Disasters. *Estimated Total Population Living Within Flood-Affected Areas, Myanmar: Flood Analysis with MODIS Terra & Aqua Data Recorded 5 May & 15 April 2008 and LandScan 2008*. TC-2008-000057-MMR, Version 1.2. May 6, 2008.

IPCC. *16 Years of Scientific Assessment in Support of the Climate Convention*. December 2004.

IPCC. *About IPCC*. www.ipcc.ch/about/index.htm.

IPCC. *Climate Change 2001: Synthesis Report Website*. www.ipcc.ch/ipccreports/tar/vol4/english/index.htm.

IPCC. *IPCC Reports Website*. www.ipcc.ch/ipccreports/assessments-reports.htm.

IPCC. *Special Report on Emissions Scenarios*. N. Nakicenovic and R. Swart (Eds.). www.grida.no/climate/ipcc/emission/.

Kaiser, Ralf I. "Experimental Investigation on the Formation of Carbon-Bearing Molecules in the Interstellar Medium via Neutral-Neutral Reactions." *Chem. Rev.,* 102(5), 1309-1358 (May 2002).

Kopp, Raymond J. and William J. Pizer (Eds.). *Assessing U.S. Climate Policy Options: A report summarizing work at RFF as part of the inter-industry U.S. Climate Policy Forum*. Resources for the Future. November 2007. www.rff.org/Publications/Pages/CPF_AssessingUSClimate PolicyOptions.aspx.

Krump, L.R., J.F. Kasting, and R.G. Crane. *The Earth System*. 2nd Edition, Pearson Education Inc. 2004.

Letter from Donna M. Heivilin. Director of Planning and Reporting, Government Accountability Office, to the Honorable Jerry Lewis, Chairman, Subcommittee on Defense, Committee on Appropriations, House of Representatives. "Military Operations: DOD's Disaster Relief Assistance in Response to Hurricane Mitch." March 29, 1999.

Lichtenstein, Michael David. *Regulating Climate Change: A Summary of Federal and State Action*. The Metropolitan Corporate Counsel. April 11 2008: 30.

Lovins, Amory B. and David R. Kramer. "Hypercars, Hydrogen and the Automotive Transition." *International Journal of Vehicle Design*, Vol. 35, Nos. 1/2, 50–85 (March 23, 2004). rmi.org/images/PDFs/ Transportation/T04-01_HypercarH2AutoTrans.pdf.

Max, Arthur. *Climate Negotiators Review Key Issues in Germany*. Associated Press. June 12, 2008.

McGray, Heather, Anne Hawmill, Rob Bradley. *Weathering the Storm: Options for Framing Adaptation and Development*. A Report of the World Resources Institute (WRI). Contributing editors List Schipper and Jo-Ellen Parry. Washington, DC, 2007. www.wri.org/publication/weathering-the-storm.

Natural Resources Canada. *Climate Change Impacts and Adaptation Program*. November 2008. adaptation.nrcan.gc.ca/index_e.php.

Natural Resources Canada. *Climate Change Impacts and Adaptation: A Canadian Perspective*. Donald S. Lemmen and Fiona J. Warren (eds.). Climate Change Impacts and Adaptation Directorate. 2004. adaptation.nrcan.gc.ca/perspective/index_e.php.

NOAA. National Marine Sanctuaries Website**.** *Frequently Asked Questions*. August 03, 2006. sanctuaries.noaa.gov/about/faqs/welcome.html.

NOAA. *What is an El Niño?* www.pmel.noaa.gov/tao/elnino/el-nino-story.html.

Nordhaus, William. *The Challenge of Global Warming: Economic Models and Environmental Policy*. Yale University, New Haven, CT. July 24, 2007. nordhaus.econ.yale.edu/.

Parker, Larry. *Climate Change: Design Approaches for a Greenhouse Gas Reduction Program*. Congressional Research Service Report to Congress. January 16, 2007.

Parry, Ian W. H. *Should we Abandon Cap-and-Trade in Favor of a CO_2 Tax? A Weathervane Commentary*. Resources for the Future. March 23, 2007. www.rff.org/Publications/Pages/PublicationDetails.aspx?PublicationID=11421.

Pew Center on Global Climate Change. *Adaptation Planning - What U.S. States and Localities are Doing*. November 2007 (updated September 2008). www.pewclimate.org/docUploads/State-Adapation-Planning-090908.pdf.

Pew Center on Global Climate Change. *Adaptation to Climate Change: International Policy Options, Advancing the International Effort Against Climate Change*. Ian Burton, Elliot Diringer and Joel Smith. Arlington, VA. November 2006. www.pewtrusts.org/uploadedFiles/wwwpew trustsorg/Reports/Global_warming/pew_climate_adaptation1106.pdf.

Pew Center on Global Climate Change. *Adapting to Climate Change: A Business Approach, Markets and Business Strategy*. Frances G. Sussman and Randall J. Freed. April 2008. www.pewclimate.org/docUploads/Business-Adaptation.pdf.

Pew Center on Global Climate Change. *Coping with Global Climate Change: The Role of Adaptation in the United States, Environment.* William E. Easterling, Brian H. Hurd and Joel B. Smith. June 2004. www.pewclimate.org/global-warming-in-depth/all_reports/adaptation/.

Pew Center on Global Climate Change. *Getting Ahead of the Curve: Corporate Strategies That Address Climate Change*. Prepared for the Pew Center on Global Climate Change by Hoffman, Andrew J., Douglas Glancy, Michael Horn, Scott Pryor, Mark Shahinian, and Groegory Shopoff, The University of Maryland. October 2006. www.pewclimate.org/docUploads/Synthesis_Report_CorpStrategies.pdf.

Pew Center on Global Climate Change. *Regional Impacts of Climate Change: Four Case Studies in the United States*. Kristie L. Ebi, Gerald A. Meehl, Dominique Bachelet, Robert R. Twilley, and Donald F. Boesch. December 2007. www.pewclimate.org/regional_impacts.

Pew Center on Global Climate Change. *The European Union Emissions Trading Scheme (EU-ETS): Insights and Opportunities.* www.pewclimate.org/docUploads/EU%2DETS%20White%20Paper %2Epdf.

Pielke, Roger A. Jr. Statement to the House Committee on Science and Technology of the United States House of Representatives, Hearing on "The State of Climate Change Science 2007: The Findings of the Fourth Assessment Report by the Intergovernmental Panel on Climate Change (IPCC), Working Group III: Mitigation of Climate Change." May 16, 2007. sciencepolicy.colorado.edu/admin/publication_files/resource-2521-2007.17.pdf

Province of British Columbia. *B.C. Government Operations: B.C. Greenhouse Gas Progress Report, 2001-2004*. May 2006.

Province of British Columbia. *Weather, Climate and the Future: B.C.'s Plan*. Victoria, B.C.: Ministry of Water, Land and Air Protection. December 2004.

Rahmstorf, S. "Thermohaline Ocean Circulation" In: *Encyclopedia of Quaternary Sciences.* Edited by S.A. Elias Elsevier. Amsterdam. 2006.

Seager, R., *et al.* "Model Projections of an Imminent Transition to a More Arid Climate in Southwestern North America." *Science*, April 5, 2007, Page 1 / 10.1126. www.ldgo.columbia.edu/cicar/documents/Sc_Express_Model_Predictionsv2.pdf.

Selin, Henrik and Stacy D. VanDeever. "Political Science and Prediction: What's Next for U.S. Climate Change Policy?" *Review of Policy Research*, 24, no.1 (2007): 1-27.

Shindell, Drew T., Greg Faluvegi, Ron L. Miller, Gavin A. Schmidt, James E. Hansen, and Shan Sun. "Solar and anthropogenic forcing of tropical hydrology." *Geophysical Research Letters*, Vol. 33, L24706, 2006. pubs.giss.nasa.gov/docs/2006/2006_Shindell_etal_4.pdf.

Socolow, Robert, Roberta Hotinski, Jeffery B. Greenblatt and Stephen Pacala. "Solving the Climate Problem: Technologies Available to Curb CO2 Emissions." *Environment*, 46, no. 10 (2004): 8-19.

Specter, Michael. "Big Foot." *The New Yorker*, February 25, 2008: 44-53.

U.S. Office for Personnel Management. *Appointment of Affidavits.* https://www.opm.gov/forms/pdf_fill/SF61.pdf.

UNFCCC. *Climate Change: Impacts, Vulnerabilities and Adaptation in Developing Counties*. Bonn. 2007. unfccc.int/files/essential_background/background_publications_htmlpdf/application/txt/pub_07_impacts.pdf.

U.S. Bureau of Reclamation. *Bureau of Reclamation - About Us*. March 19, 2008. www.usbr.gov/main/about/.

USAID. "Defining Results-Oriented Assistance." *USAID Sourcebook*. October 2002. www.usaid.gov/pubs/sourcebook/usgov/define.html.

USAID. *Adapting to Climate Variability and Change: A Guidance Manual for Development Planning*. Washington, DC. August 2007. www.usaid.gov/our_work/environment/climate/docs/reports/cc_vamanual.pdf.

Weller, Gunter, et al. "Summary and Synthesis of the Arctic Climate Impact Assessment." *Arctic Climate Impact Assessment.* ACIA. pp. 989-1020. Cambridge University Press, New York, NY, USA, ISBN 0-521-86509-3, 2005. www.acia.uaf.edu/pages/scientific.html.

White House. Press Briefing Via Conference Call by Senior Administration Officials on IPCC Report. November 17, 2007. From Valencia, Spain, [participants Jim Connaughton, Chairman of the White House Council on Environmental Quality, Dr. Sharon Hays, U.S. Delegation At IPCC Meeting, Dr. Harlan Watson, Senior Climate Negotiator, Department of State]. www.whitehouse.gov/news/releases/2007/11/20071116-21.html.

White House. *Taking Additional Action to Confront Climate Change*. White House Fact Sheet. April 16, 2008. www.whitehouse.gov/news/releases/2008/04/20080416-7.html.

Witoelar, Rachmat. Address to Closing Plenary, UN Climate Change Conference, The Bali Roadmap. December 15, 2007.

World Meteorological Organization (WMO). *Statement on Tropical Cyclones and Climate Changes*. WMO International Workshop on Tropical Cyclones, IWTC-6, San Jose, Costa Rica. November 2006. www.wmo.ch/pages/prog/arep/tmrp/documents/iwtc_statement.pdf.

Appendix B
Figure and Table Sources

FIGURES

Figure 1-1. *Sources of U.S. GHG Emissions (in Gt CO_2-eq)*

Redrafted from DOE, EIA, *Emissions of Greenhouse Gases in the United States,* DOE/EIA-0573, 2006 (Washington, DC: DOE, November 28, 2007), www.eia.doe.gov/oiaf/1605/ggrpt/.

Figure 1-2. *Global Carbon Emissions from Fossil Fuel Burning, Cement Manufacture, and Gas Flaring, 1800–2004*

Adapted from G. Marland, T.A. Boden, and R.J. Andres. "Global CO_2 Emissions from Fossil Fuel Burning, Cement Manufacture, and Gas Flaring, 1751–2004," in *Trends: A Compendium of Data on Global Change*, CDIAC, cdiac.esd.ornl.gov/ftp/ndp030/region.1751_2004.ems.

Please note that "China" also includes data from Democratic People's Republic of Korea (North Korea), Democratic Republic of Vietnam, French Indo-China, Mongolia, Republic of South Vietnam, and Vietnam. "Asia" includes Afghanistan, Bangladesh, Bhutan, Brunei (Darussalam), Cambodia, East Timor (included with Indonesia), East and West Pakistan, Federation of Malaya-Singapore, Hong Kong, India, Indonesia, Lao People's Democratic Republic, Macau, Malaysia, Maldives, Myanmar, Nepal, Pakistan, Penisular Malaysia, Philippines, Republic of Korea, Sabah, Sarawak, Singapore, Sri Lanka, Taiwan, and Thailand. Meanwhile, the Russian Federation and USSR are included in "Europe." These categorizations are from Marland, Boden, and Andres.

Figure 1-3. *Long-Term Changes in CO_2 and Temperature*

Adapted from EPA, *Climate Change Images*, www.epa.gov/climatechange/science/images/co2-temp.gif.

Figure 1-4. *Climate Change Policy Management Structure*

Adapted from NOAA, National Climatic Data Center, *Vision Figure 2*, www.climatescience.gov/Library/stratplan2003/vision/VisionFig2.jpg.

Figure 2-1. *Carbon Fluxes and Reservoirs in the Global Carbon Cycle*

Adapted from the NASA Earth Observatory, *The Carbon Cycle: The Human Role*, 2001, earthobservatory.nasa.gov/Library/CarbonCycle/carbon_cycle4.html.

Figure 2-2. *North American Carbon Sources and Sinks*

Adapted from U.S. Climate Change Science Program and the Subcommittee on Global Change Research, The First State of the Carbon Cycle Report (SOCCR): The North American Carbon Budget and Implications for the

Global Carbon Cycle, Synthesis and Assessment Product 2.2, [King, A.W., et al. (Eds.)], NOAA, National Climatic Data Center, Asheville, NC, p. 5, Figure ES.1, www.climatescience.gov/Library/sap/sap2-2/final-report/default.htm. Used by permission of the CCSP.

Figure 2-3. *Energy Balance of Earth (in W/m^2)*

Adapted from *Climate Change 2001: Synthesis Report. A Contribution of Working Groups I, II and III to the Third Assessment Report of the Intergovernmental Panel on Climate Change.* Figure 2-2. Cambridge University Press, Cambridge, United Kingdom, and New York, NY, USA, www.ipcc.ch/ipccreports/tar/vol4/english/index.htm. Used by permission of the IPCC.

Figure 2-4. *Results of Global Temperature Change Experiments*

Adapted from "Technical Summary" in *Climate Change 2007: The Physical Science Basis. Working Group I Contribution to the Fourth Assessment Report of the Intergovernmental Panel on Climate Change*, Figure TS.23, Cambridge University Press, Cambridge, United Kingdom, and New York, NY, USA, www.ipcc.ch/pdf/assessment-report/ar4/wg1/ar4-wg1-ts.pdf. Used by permission of the IPCC.

Figure 2-5. *Climate System Components, Processes, and Interactions*

Adapted from "Frequently Asked Question 1.2: What is the Relationship Between Climate Change and Weather?" in *Climate Change 2007: The Physical Science Basis. Working Group I Contribution to the Fourth Assessment Report of the Intergovernmental Panel on Climate Change*, FAQ 1.2, Figure 1, Cambridge University Press, Cambridge, United Kingdom, and New York, NY, USA, www.ipcc.ch/graphics/gr-ar4-wg1.htm. Used by permission of the IPCC.

Figure 2-6. *Thermohaline Circulation of Global Ocean System*

Adapted from *Climate Change 2001: Synthesis Report. A Contribution of Working Groups I, II and III to the Third Assessment Report of the Intergovernmental Panel on Climate Change.* Figure 4-2. Cambridge University Press, Cambridge, United Kingdom, and New York, NY, USA, www.ipcc.ch/ipccreports/tar/vol4/english/index.htm. Used by permission of the IPCC.

Figure 2-7. *Past Global Average Sea Level*

Adapted from "Frequently Asked Question 5.1: Is Sea Level Rising?" from *Climate Change 2007: The Physical Science Basis. Working Group I Contribution to the Fourth Assessment Report of the Intergovernmental Panel on Climate Change*, FAQ 5.1, Figure 1, Cambridge University Press, Cambridge, United Kingdom, and New York, NY, USA, ipcc-wg1.ucar.edu/wg1/FAQ/wg1_faq-5.1.html. Used by permission of the IPCC.

Figure 2-8. *SRES Qualitative Storylines for Modeling Scenarios*

Adapted from the United Nations Environmental Programme (UNEP)/GRID-Arendal. 2005. *Vital Climate Change Graphics*,

UNEP/GRID-Arendal, Arendal, Norway, Nairobi, Kenya, ISBN: 82-7701-031-1, p. 6, http://www.grida.no/_res/site/file/publications/vital-climate_change_update.pdf. Used by permission of GRID-Arendal.

Figure 2-9. *Global GHG Emissions for 2000 and Projected Emissions for 2030 using SRES and Post-SRES Scenarios*

Adapted from "Technical Summary" in *Climate Change 2007: Mitigation of Climate Change. Working Group III Contribution to the Fourth Assessment Report of the Intergovernmental Panel on Climate Change*, Figure TS.5, Cambridge University Press, Cambridge, United Kingdom, and New York, NY, USA, www.ipcc.ch/pdf/assessment-report/ar4/wg3/ar4-wg3-ts.pdf. Used by permission of the IPCC.

Figure 2-10. *Global Averages of Surface Warming for SRES Scenarios*

Adapted from "Summary for Policymakers" In *Climate Change 2007: The Physical Science Basis. Working Group I Contribution to the Fourth Assessment Report of the Intergovernmental Panel on Climate Change.* Figure SPM-5, Cambridge University Press, Cambridge, United Kingdom, and New York, NY, USA, www.ipcc.ch/pdf/assessment-report/ar4/wg1/ar4-wg1-spm.pdf. Used by permission of the IPCC.

Figure 2-11. *Evolution of Climate Models*

Adapted from "Technical Summary" in *Climate Change 2001: Synthesis Report. A Contribution of Working Groups I, II and III to the Third Assessment Report of the Intergovernmental Panel on Climate Change.* Synthesis Report Questions, Box 3, Figure 1, Cambridge University Press, Cambridge, United Kingdom, and New York, NY, USA, www.ipcc.ch/ipccreports/tar/vol4/english/wg1figts-box3.htm. Used by permission of the IPCC.

Figure 3-1. *Comparative Effects on Emissions (to 2050) of Proposals in 110th Congress*

World Resources Institute, *Comparison of Legislative Climate Change Targets*, John Larsen and Robert Heilmayr, September 9, 2008. Used by permission of the World Resources Institute.

Figure 3-2. *December 2008 Futures Prices for One Allowance Unit (1 Ton CO_2) in January–July 2008*

European Climate Exchange, *Historical data—ECX EUA Futures Contract*, 2008, www.europeanclimateexchange.com/default_flash.asp. Used by permission of the European Climate Exchange.

Figure 3-3. *DOE Projection of Worldwide Carbon Emissions under Business-as-Usual Assumptions*

DOE EIA, *International Energy Outlook 2008, Highlights*, www.eia.doe.gov/oiaf/ieo/pdf/highlights.pdf.

Figure 3-4. *GHG Emission Stabilization Wedges*

Adapted from *Stabilization Wedges*, www.princeton.edu/wedges/presentation_resources/and R. H. Socolow, *Introduction to the Wedges*, www.princeton.edu/~cmi/resources/stabwedge.htm, both by the Carbon Mitigation Initiative, Princeton University. Used by permission of the Carbon Mitigation Initiative, Princeton University.

Figure 4-1. *GHG Scopes*

Redrafted from World Resources Institute and World Business Council for Sustainable Development, *Greenhouse Gas Protocol, A Corporate Accounting and Reporting Standard, Revised Edition*, Geneva, Switzerland, and Washington, DC, www.ghgprotocol.org/files/ghg-protocol-revised.pdf. Used by permission of the World Resources Institute.

You Have the Power graphic

DOE, FEMP, *You Have the Power Campaign Materials*, August 15, 2008, www1.eere.energy.gov/femp/services/yhtp/campaign_materials.html.

Figure 4-2. *U.S. Commercial Building Energy Use in 2005*

Redrafted from "Residential and commercial buildings" in *Climate Change 2007: Mitigation of Climate Change. Working Group III Contribution to the Fourth Assessment Report of the Intergovernmental Panel on Climate Change*. Figure 6.3. Cambridge University Press, Cambridge, United Kingdom, and New York, NY, USA, www.ipcc.ch/pdf/assessment-report/ar4/wg3/ar4-wg3-chapter6.pdf. Used by permission of the IPCC.

Figure 4-3. *GHG Reduction Opportunities Widely Distributed—2030 Mid-Range Case*

Adapted from J. Creyts, A. Derkach, S. Nyquist, K. Ostrowski, and J. Stephenson, *Reducing U.S. Greenhouse Gas Emissions: How Much at What Cost?*, Executive Report of the U.S. Greenhouse Gas Abatement Mapping Initiative, McKinsey & Company, December 2007, www.international-alert.org/pdf/A_Climate_Of_Conflict.pdf. Used by permission of McKinsey & Company.

Figure 5-1. *Increase in Percentage of Very Warm Nights (top 10 percent) in North America*

Redrafted from CCSP, 2008: *Weather and Climate Extremes in a Changing Climate. Regions of Focus: North America, Hawaii, Caribbean, and U.S. Pacific Islands.* A Report by the U.S. Climate Change Science Program and the Subcommittee on Global Change Research. [Thomas R. Karl, Gerald A. Meehl, Christopher D. Miller, Susan J. Hassol, Anne M. Waple, and William L. Murray (eds.)]. Department of Commerce, NOAA's National Climatic Data Center, Washington, DC, USA, 164 pp., Figure 3.1, www.climatescience.gov/Library/sap/sap3-3/final-report/default.htm. Used by permission of the CCSP.

Figure 5-2. *Periods of Heavy Rainfall Lasting at Least 90 Days in United States, 1895–2005*

Redrafted from CCSP, 2008: *Weather and Climate Extremes in a Changing Climate. Regions of Focus: North America, Hawaii, Caribbean, and U.S. Pacific Islands.* A Report by the U.S. Climate Change Science Program and the Subcommittee on Global Change Research. [Thomas R. Karl, Gerald A. Meehl, Christopher D. Miller, Susan J. Hassol, Anne M. Waple, and William L. Murray (eds.)]. Department of Commerce, NOAA's National Climatic Data Center, Washington, DC, USA, 164 pp., Figure 2.9, www.climatescience.gov/Library/sap/sap3-3/final-report/default.htm. Used by permission of the CCSP.

Figure 5-3. *Increase in Percentage of Heavy Daily Precipitation in North America*

Redrafted from CCSP, 2008: *Weather and Climate Extremes in a Changing Climate. Regions of Focus: North America, Hawaii, Caribbean, and U.S. Pacific Islands.* A Report by the U.S. Climate Change Science Program and the Subcommittee on Global Change Research. [Thomas R. Karl, Gerald A. Meehl, Christopher D. Miller, Susan J. Hassol, Anne M. Waple, and William L. Murray (eds.)]. Department of Commerce, NOAA's National Climatic Data Center, Washington, DC, USA, 164 pp., Figure 3.1, www.climatescience.gov/Library/sap/sap3-3/final-report/default.htm. Used by permission of the CCSP.

Figure 5-4. *Projected Increases in Very Heavy Rainfall Events in North America*

Redrafted from CCSP, 2008: *Weather and Climate Extremes in a Changing Climate. Regions of Focus: North America, Hawaii, Caribbean, and U.S. Pacific Islands.* A Report by the U.S. Climate Change Science Program and the Subcommittee on Global Change Research. [Thomas R. Karl, Gerald A. Meehl, Christopher D. Miller, Susan J. Hassol, Anne M. Waple, and William L. Murray (eds.)]. Department of Commerce, NOAA's National Climatic Data Center, Washington, DC, USA, 164 pp., Figure 3.4, www.climatescience.gov/Library/sap/sap3-3/final-report/default.htm. Used by permission of the CCSP.

We caution the reader not to draw site-specific conclusions from this figure. The figure has been redrafted from the original figure, not from the original data, and we were not able to replicate grid cell locations and patterns exactly. Please use this figure to consider general patterns and refer to the original figure or original data for more specific information.

Figure 5-5. *Average Number of Hurricanes/Tropical Storms and Sea Surface Temperatures for North Atlantic over Past Century*

Please note, the blue line represents the 9-year running average number of hurricanes and tropical storms. The red line represents the 9-year smoothed sea surface temaperature.

Adapted from CCSP, 2008, *Weather and Climate Extremes in a Changing Climate. Regions of Focus: North America, Hawaii, Caribbean, and U.S. Pacific Islands*, A Report by the U.S. Climate Change Science Program and the Subcommittee on Global Change Research. [Thomas R. Karl, Gerald A. Meehl, Christopher D. Miller, Susan J. Hassol, Anne M. Waple, and William L. Murray (eds.)], Department Commerce, NOAA's National Climatic Data Center, Washington, DC, USA, 164 pp., Figure 2-15, www.climatescience.gov/Library/sap/sap3-3/final-report/default.htm.

Figure 5-6. *Vegetation of Arctic: Current Conditions and Projected Changes under S92a Scenario (2090–2100)*

Adapted from "Technical Summary" in *Climate Change 2007: Impacts, Adaptation and Vulnerability. Working Group II Contribution to the Fourth Assessment Report of the Intergovernmental Panel on Climate Change*, Figure TS.16, Cambridge University Press, Cambridge, United Kingdom, and New York, NY, USA, www.ipcc.ch/pdf/assessment-report/ar4/wg2/ar4-wg2-ts.pdf. Used by permission of the IPCC.

Please note, IS92a is a pre-SRES emission scenario that assumes a 1 percent rise in CO_2 per year from 1990 levels. IPCC created this figure using several data sources; the sea ice projection is for the 2080–2100 period, while other aspects of the figure refer to the 2090–2100 period mentioned in the caption.

Figure 5-7. *September Arctic Ice Extent from 1979 to 2007*

Redrafted from CCSP, 2008, *Our Changing Planet, the U.S. Climate Change Science Program (CCSP) for Fiscal Year 2009,* A Report by the U.S. Climate Change Science Program and the Subcommittee on Global Change Research, A Supplement to the President's Budget for FY 2009, www.usgcrp.gov/usgcrp/Library/ocp2009/default.htm. Used by permission of the CCSP.

.Figure 5-8. *Spatial Distribution of Arctic Sea Ice in September 2007*

Adapted from CCSP, 2008. *Our Changing Planet, the US Climate Change Science Program (CCSP) for Fiscal Year 2009*, A Report by the U.S. Climate Change Science Program and the Subcommittee on Global Change Research, A Supplement to the President's Budget for FY 2009, www.usgcrp.gov/usgcrp/Library/ocp2009/default.htm. Used by permission of the CCSP.

Figure 6-1. *Examples of Global Impacts Projected for Changes in Climate*

The number following each impact indicates its source, as follows: (1) AR4 WG2, Table 19.1; (2) AR4 WG2, Table 19.1; (3) AR4 WG2, Section 10.4; (4) AR4 WG2, Table 19.1; (5) AR4 WG2, Table 19.1; (6) AR4 WG2, Section 19.3; (7) AR4 WG2, Sections 4.6, 6.4; (8) AR4 WG2, Table 19.1; (9) AR4 WG2, Box TS.5 (very likely); (10) AR4 WG2, Sections 6.3, 6.4; (11) AR4 WG2, Table 19.1 (12) AR4 SYR SPM, Figure SPM.7, AR4 WG2, Box TS.5; (13) AR4 WG2, Table 19.1; (14) AR4 WG2, Table 19.1; (15)

AR4 WG2, Section 19.3. The source for scenario component is AR4 WG2, Figure TS.3.

AR4 WG2 is *Climate Change 2007: Impacts, Adaptation and Vulnerability. Working Group II Contribution to the Fourth Assessment Report of the Intergovernmental Panel on Climate Change*, Cambridge University Press, Cambridge, United Kingdom, and New York, NY, USA, www.ipcc.ch/ipccreports/ar4-wg2.htm.

AR4 SYR SPM is "Summary for Policymakers" in *Climate Change 2007: Synthesis Report. Contribution of Working Groups I, II and III to the Fourth Assessment Report of the Intergovernmental Panel on Climate Change*, Geneva, Switzerland, www.ipcc.ch/pdf/assessment-report/ar4/syr/ar4_syr_spm.pdf.

Figure 6-2. *Number of People at Risk Globally from Coastal Flooding* (Annually in Millions)

Redrafted from "Technical Summary" in *Climate Change 2007: Impacts, Adaptation, and Vulnerability. Working Group II Contribution to the Fourth Assessment Report of the Intergovernmental Panel on Climate Change*, Figure TS.18, Cambridge University Press, Cambridge, United Kingdom, and New York, NY, USA, www.ipcc.ch/pdf/assessment-report/ar4/wg2/ar4-wg2-ts.pdf. Used by permission of the IPCC.

Figure 6-3. *Decadal Average Hurricane Total Dissipated Energy (PDI), Loss of Life, and Inflation-Adjusted Economic Damages (in terms of $000) from Hurricanes Making Landfall in the Continental United States*

Redrafted from "Technical Summary" in *Climate Change 2007: Impacts, Adaptation, and Vulnerability. Working Group II Contribution to the Fourth Assessment Report of the Intergovernmental Panel on Climate Change*, Figures TS.15, Cambridge University Press, Cambridge, United Kingdom, and New York, NY, USA, www.ipcc.ch/pdf/assessment-report/ar4/wg2/ar4-wg2-ts.pdf. Used by permission of the IPCC.

Figure 7-1. *Relative Vulnerability of Coastal Deltas (IPCC)*

Redrafted from "Technical Summary" in *Climate Change 2007: Impacts, Adaptation, and Vulnerability. Working Group II Contribution to the Fourth Assessment Report of the Intergovernmental Panel on Climate Change*, Figure TS.8, Cambridge University Press, Cambridge, United Kingdom, and New York, NY, USA, www.ipcc.ch/pdf/assessment-report/ar4/wg2/ar4-wg2-ts.pdf. Used by permission of the IPCC.

Figure 7-2. *Projected Relative Percentage Changes in Precipitation by 2100*

Adapted from "Summary for Policymakers" in *Climate Change 2007: The Physical Science Basis. Working Group I Contribution to the Fourth Assessment Report of the Intergovernmental Panel on Climate Change*, Table SPM-7, Cambridge University Press, Cambridge, United Kingdom, and New York, NY, USA, www.ipcc.ch/pdf/assessment-report/ar4/wg1/ar4-wg1-spm.pdf. Used by permission of the IPCC.

Figure 7-3. *Projected Risk of Armed Conflict as Consequence of Climate Change*

Redrafted from Dan Smith and Janani Vivekanda, "A Climate of Conflict: The Links Between Climate Change, Peace and War." *International Alert*, November 2007, www.international-alert.org/pdf/A_Climate_Of_Conflict.pdf. Used by permission of International Alert.

Figure 7-4. *Office of Foreign Disaster Assistance Disaster Funding FY06 by Region*

Adapted from USAID, Office of U.S. Foreign Disaster Assistance, *Annual Report for Fiscal Year 2006*, 2007, www.usaid.gov/our_work/humanitarian_assistance/disaster_assistance/publications/annual_reports/pdf/AR2006.pdf.

TABLES

Table 1-1. *GHG Characteristics*

Adapted from T.J. Blasing and K. Smith, *Recent Greenhouse Gas Concentrations*, Carbon Dioxide Information Analysis Center (CDIAC), July 2006, cdiac.ornl.gov/pns/current_ghg.html.

Table 1-2. *Federal Agency Reporting on Climate Change Actions and Plans*

Data from DOE, FY07 Federal Energy Management Program, www1.eere.energy.gov/femp/index.html.

Table 1-3. *USGCRP Agency Participation*

USGCRP, *U.S. Government Agencies Participating in the USGCRP*, www.usgcrp.gov/usgcrp/agencies/.

Table 1-4. *GAO Reports Related to Climate Change*

GAO, *Welcome to GAO*, www.gao.gov; search for climate change.

Table 2-1. *Observed Rate of SLR and Estimated Contributions*

"Summary for Policymakers." In *Climate Change 2007: The Physical Science Basis. Working Group I Contribution to the Fourth Assessment Report of the Intergovernmental Panel on Climate Change*, Table SPM-1, Cambridge University Press, Cambridge, United Kingdom, and New York, NY, USA, www.ipcc.ch/pdf/assessment-report/ar4/wg1/ar4-wg1-spm.pdf.

Table 3-1. *U.S. Per Capita Covered GHGs*

Lieberman-Warner Climate Security Act of 2007, Senate Bill 2191, proposed by Senators Joe Lieberman (I-CT) and John Warner (R-VA), introduced October 18, 2007.

Table 4-1. *Four Standards of GHG Protocol*

Drawn from *Greenhouse Gas Protocol, A Corporate Accounting and Reporting Standard, Revised Edition*, Geneva, Switzerland, and Washington DC, www.ghgprotocol.org/files/ghg-protocol-revised.pdf.

Table 4-2. *Steps for Developing a GHG Management Program*

Adapted from Ceres and the Investor Network on Climate Risk, *Managing the Risks and Opportunities of Climate Change: A Practical Toolkit for Corporate Leaders*, 2006, p. 23, 216.235.201.250/netcommunity/Document.Doc?id=92, and Pew Center, *Getting Ahead of the Curve: Corporate Strategies That Address Climate Change*, prepared for the Pew Center on Global Climate Change, 2006, www.pewclimate.org/docUploads/Synthesis_Report_CorpStrategies.pdf.

Table 4-3. *Comparison of Fuel Mileage and GHG Emissions*

Source for fuel data: Norman Brinkman, Michael Wang, Trudy Weber, and Thomas Darlington, *Well-to Wheels Analysis of Advanced Fuel/Vehicle Systems—A North American Study of Energy Use, Greenhouse Gas Emissions, and Criteria Pollutant Emissions,* General Motors Corporation, Argonne National Laboratory, and Air Improvement Resource, Inc., May 2005, www.transportation.anl.gov/pdfs/TA/339.pdf. Source for public transit data: M. J. Bradley & Associates, *Comparison of Energy Use and CO_2 Emissions From Different Transportation Modes*, for the American Bus Association, Manchester, NH, May 2007, www.buses.org/files/ComparativeEnergy.pdf.

Table 4-4. *Light Bulb Replacement*

EPA, "Tools and resources for lighting retrofit projects," *eeBuildings*, February 4, 2008, www.epa.gov/eebuildings/lighting/detail/index.html.

Table 4-5. *Opportunities for Energy Consumption Savings*

Extracted from The Carbon Trust, *Local authorities—Saving energy in local authority buildings*, Sector Overview CTV028, London, UK, June 2007, www.carbontrust.co.uk/carbon/publicsector/la/.

Table 4-6. *Life-Cycle Carbon Emissions from Office Paper*

DOE, E.O. Lawrence Berkeley National Laboratory, University of California, Environmental Energy Technologies Division (EETD), "Facts about U.S. Paper Use," *Cutting Paper*, eetd.lbl.gov/paper/ideas/html/uspaperuse.htm.

Table 4-7. *Global Economic Mitigation Potential in 2030*

Adapted from "Technical Summary" in *Climate Change 2007: Mitigation of Climate Change. Working Group III Contribution to the Fourth Assessment Report of the Intergovernmental Panel on Climate Change*, Tables TS.15 and TS.16, Cambridge University Press, Cambridge, United Kingdom and New York, NY, USA, www.ipcc.ch/pdf/assessment-report/ar4/wg3/ar4-wg3-ts.pdf.

Table 4-8. *Examples of Climate Change-Related Technology Research and Development Programs*

Information drawn from individual agency websites and the Climate Change Technology Program website, www.climatetechnology.gov/.

Table 5-1. *IPCC Consensus on Future Climate Change Impacts with High Likelihood*

Climate Change 2007: Impacts, Adaptation and Vulnerability. Working Group II Contribution to the Fourth Assessment Report of the Intergovernmental Panel on Climate Change, Chapter 8, Cambridge University Press, Cambridge, United Kingdom, and New York, NY, USA, 976 pp., www.ipcc.ch/ipccreports/ar4-wg2.htm.

Table 5-2. *Sample Impacts on Specific Agency Operations*

Information drawn from individual agency websites.

Table 5-3. *Population Projections for United States and Its Coastal Areas*

Adapted from U.S. Census Bureau, *Annual Projections of the Total Resident Population as of July 1: Middle, Lowest, Highest, and Zero International Migration Series, 1999 to 2100, NP-T1*, January, 2000, www.census.gov/population/projections/nation/summary/np-t1.pdf.

Table 5-4. *Land and Economic Loss from Sea-Level Rise*

Adapted from James G. Titius, Richard A. Park, Stephen P. Leatherman, J. Richard Weggel, Michael S. Greene, Paul W. Mausel, Scott Brown, Gary Gaunt, Manjit Trehan, and Gary Yohe, "Greenhouse Effect and Sea Level Rise: The Cost of Holding Back the Sea," *Coastal Management* (1991), Volume 19, 171–204, www.epa.gov/climatechange/effects/coastal/slrmaps_cost_of_holding.html#abstract.

Estimated economic losses displayed in $ billion, converted from 1988 to 2008 estimated dollars using NASA, "Gross Domestic Product Deflator Inflation Calculator," *Cost Estimating Web Site*, cost.jsc.nasa.gov/inflateGDP.html.

Table 6-1. *Climate Change Adaptation Categories*

B. Mehdi, Charles Mrena, and Al Douglas, *Adapting to Climate Change: An Introduction for Canadian Municipalities*, Canadian Climate Impacts and Adaptation Research Network (C-CIARN), 32 pp., February 2006, www.adaptation.nrcan.gc.ca/municipalities_e.php.

Table 6-2. *Steps in Vulnerability Assessment*

Adapted from D. S. Lemmen and F. J. Warren, eds., *Climate Change Impacts and Adaptation: A Canadian Perspective* (Natural Resources Canada, 2004), Figure 1, 174 pp., adaptation.nrcan.gc.ca/perspective/pdf/report_e.pdf.

Table 7-1. *Federal Agency Adaptive Response to Climate Change*

Information drawn from individual agency websites.

Table 7-2. *IPCC Summary of Climate-Related Impacts in Coastal Zones*

"Coastal systems and low-lying areas" in *Climate Change 2007: Impacts, Adaptation and Vulnerability. Working Group II Contribution to the Fourth Assessment Report of the Intergovernmental Panel on Climate Change*, Table 6.4, Cambridge University Press, Cambridge, United Kingdom, and

New York, NY, USA, www.ipcc.ch/pdf/assessment-report/ar4/wg2/ar4-wg2-chapter6.pdf.

Table 7-3. *Agency Response to Cyclone Nargis*

Information drawn from individual agency websites.

Appendix C
IPCC Findings

As described in Chapter 1, we have created a searchable database of the more than 2,700 findings in the IPCC's AR4 (fourth assessment report), each of which is accompanied by some measure of uncertainty, ranging from low to high. In this guide, we only include findings in which the IPCC expressed a high level of scientific understanding. These are the IPCC findings with (1) high agreement and much evidence (qualitatively most reliable); (2) very high confidence (quantitatively most reliable, at least a 9 in 10 chance of being correct); (3a) very likely (at least 90 percent probability); or (3b) very unlikely (less than 10 percent probability).

These choices constitute the facts of climate change best supported by the world community at this time. They are highlighted in the purple boxes throughout the guide, as well as in Table 5-1 (see "Figure and Table Sources"), and their use and selection is discussed in Chapter 1. Below, we show each IPCC finding, indicating the page of citation in this guide, initial words, and source in the AR4, including section number when appropriate (see below for full citations).

Chapter 1

Page	Finding text	Source
1	*Most of the observed increase...*	AR4 SYR, SPM 2.4
4	*Warming of the climate system...*	AR4 SYR, SPM 1
5	*It is very likely that over...*	AR4 SYR, SPM 1.1
10	*There is high confidence...2004.*	AR4 WG3, SPM 3

Chapter 2

Page	Finding text	Source
22	*Continued GHG emissions...*	AR4 SYR, SPM 3.2.1
22	*It is very likely that the observed...*	AR4 SYR, SPM 2.2
27	*In order to stabilize the...*	AR4 WG3, SPM 15
28	*The combined radiative forcing due...*	AR4 SYR, 2.2
29	*There is very high confidence that...* W/m^2.	AR4 SYR, 2.2
30	*The observed pattern of tropospheric...*	AR4 SYR, 2.4
32	*The observed widespread warming...*	AR4 SYR, 3.4
35	*New data since the IPCC...*	AR4 WG1, SPM
36	*Based on current model simulations...*	AR4 SYR, SPM 3.4
39	*Average Northern Hemisphere...*	AR4 SYR, SPM 1.1
40	*Human influences have very likely...*	AR4 SYR, SPM 2.4
41	*There is very high confidence, based...*	AR4 SYR, 1.2
42	*A synthesis of studies strongly...*	AR4 SYR, 2.4
45	*Progress since the IPCC Third...*	AR4 SYR, 2.3
45	*Best-estimate projections from models...*	AR4 WG1, SPM
47	*It is very unlikely that climate changes...*	AR4 WG1, SPM

CHAPTER 3

Page	Finding text	Source
51	*There is high agreement...instrument.*	AR4 SYR, 4.3
58	*There is high agreement...regulation.*	AR4 WG3, SPM 19
74	*There is high agreement...decades.*	AR4 SYR, SPM 3.1
75	*There is high agreement...nations.*	AR4 WG3, SPM 14
76	*There is high agreement...financing.*	AR4 WG3, SPM 20

CHAPTER 4

Page	Finding text	Source
83	*Both bottom-up and top-down studies...*	AR4 SYR, SPM 4.3
87	*There is high agreement...delay.*	AR4 WG3, SPM 18, 3
90	*There is high agreement...barriers.*	AR4 SYR, SPM 5.5
94	*There is high agreement...co-benefits.*	AR4 WG3, SPM 13
100	*There is high agreement...waste...*	AR4 WG3, SPM 14
107	*There is high agreement...forest...*	AR4 WG3, SPM 14
114	*While studies use different...*	AR4 SYR, SPM 4.3

CHAPTER 5

Page	Finding text	Source
123	*It is very likely that hot...*	AR4 SYR, Table 3.2
125	*North America—Cities that currently...*	AR4 SYR, SPM, 3.3.2
126	*The balance of positive and negative...*	AR4 WG2, SPM
132	*It is very likely that globally...*	AR4 SYR, 5.7
132	*In some locations and amongst...*	AR4 SYR, 5.7
133	*Impacts of climate change will vary...*	AR4 SYR, 5.7
135	*Coasts are projected to be...*	AR4 SYR, 3.3.1
136	*Corals are vulnerable to thermal...*	AR4 SYR, SPM, 5
137	*There is very high confidence...sediment.*	AR4 WG2, SPM
138	*Sea level rise is expected to...*	AR4 WG2, SPM
140	*There is very high confidence...resources.*	AR4 SYR, 3.3.1
142	*North America—Disturbances...*	AR4 WG2, SPM
142	*North America—Warming...*	AR4 SYR, SPM 3
143	*Increases in the amount of...*	AR4 SYR, SPM 3

CHAPTER 6

Page	Finding text	Source
155	*North America—There is very high ...*	AR4 SYR, SPM 3
155	*There is very high confidence...low.*	AR4 WG2, SPM
155	*There is very high confidence...plans.*	AR4 WG2, SPM
156	*Small islands, whether located in...*	AR4 WG2, SPM
159	*There is very high confidence...uncertainties.*	AR4 WG2, SPM
160	*Climate change is projected...*	AR4 WG2, SPM
161	*Impacts are very likely...*	AR4 SYR, 6.2

CHAPTER 7

Page	Finding text	Source
177	*There is high agreement...suggested.*	AR4 SYR, SPM 4.5
178	*The literature provides...*	AR4 SYR, 4.5
179	*There is high agreement...employment.*	AR4 WG3, SPM.13
181	*By the 2080s, many millions...*	AR4 SYR, 3.3.1
182	*Australia and New Zealand—By 2020...*	AR4 SYR, SPM 3
184	*Europe—There is very high...rise.*	AR4 SYR, SPM 3
184	*Europe—There is very high...2080.*	AR4 SYR, SPM 3
190	*It is very likely that climate...*	AR4 SYR, 5.8
191	*There is high agreement...mitigation...*	AR4 WG3, SPM 21

SOURCE DOCUMENTS

AR4 SYR: *Climate Change 2007: Synthesis Report. Contribution of Working Groups I, II and III to the Fourth Assessment Report of the Intergovernmental Panel on Climate Change*, Geneva, Switzerland, www.ipcc.ch/pdf/assessment-report/ar4/syr/ar4_syr_spm.pdf.

AR4 WG1, SPM: "Summary for Policymakers" in *Climate Change 2007: The Physical Science Basis. Working Group I Contribution to the Fourth Assessment Report of the Intergovernmental Panel on Climate*, Cambridge University Press, Cambridge, United Kingdom, and New York, NY, USA, www.ipcc.ch/pdf/assessment-report/ar4/wg1/ar4-wg1-spm.pdf.

AR4 WG2, SPM: "Summary for Policymakers" in *Climate Change 2007: Impacts, Adaptation and Vulnerability. Working Group II Contribution to the Fourth Assessment Report of the Intergovernmental Panel on Climate Change*, Cambridge University Press, Cambridge, United Kingdom, and New York, NY, USA, 976 pp., www.ipcc.ch/ipccreports/ar4-wg2.htm.

AR4 WG3, SPM: "Summary for Policymakers" in *Climate Change 2007: Mitigation of Climate Change. Working Group III Contribution to the Fourth Assessment Report of the Intergovernmental Panel on Climate Change*, Cambridge University Press, Cambridge, United Kingdom, and New York, NY, USA, www.ipcc.ch/pdf/assessment-report/ar4/wg3/ar4-wg3-chapter6.pdf.

Appendix D Reviewers

We are grateful to the following people, who provided expert review of this guide.

Jeremey Alcorn
Concurrent Technologies Corporation

Dana Arnold
Office of the Federal Environmental Executive

Ed Bailey
Marymount University

Molly Brown
Science Systems and Applications, Inc.

Kathleen Cardamone
The Government Accountability Office

Christina Crockett
George Mason University

Bob Curran
U.S. Global Change Research Program

Brigid Decoursey
U.S. Department of Transportation

Lisa Dilling
University of Colorado at Boulder

Emil J. Dzuray, Jr.
U.S. Postal Service

Steven Gabriel
Visiting Scholar, LMI Research Institute; University of Maryland

Bill Goran
U.S. Army Corps of Engineers

Tee Guidotti
George Washington University

Sam Higuchi
National Aeronautics and Space Administration

Doug James
National Science Foundation

Robert Kafalenos
U.S. Department of Transportation

Leah Bunce Karrer
Conservation International

Bob Krumenaker
National Park Service

Craig Lasser
U.S. Forest Service

Pedro Morales
Office of the Secretary of Defense

Dennis Ojima
Heinz Center

Roger Pielke, Jr.
University of Colorado at Boulder

Tony Prato
University of Mississippi

Rob Ritter
Department of Transportation

Chet Ropeleski
National Oceanic and Atmospheric Administration

Michael Savonis
Department of Transportation

Robert Shaw
Texas A&M Institute of Renewable Natural Resources

Peter Schultz, *U.S. Global Change Research Program*

Arthur Jack "A.J." Singletary
Department of Transportation

Steven L Sosson
Department of the Treasury

Alan Strasser
The Department of Transportation

Eric Toman
The Ohio State University

Tom Triscari, *Distinguished Visiting Scholar, LMI Research Institute; Rensselaer Polytechnic Institute*

Charles Wilson
U.S. Navy

Gary Willson
National Park Service

Appendix E
Glossary

Adaptation. Any adjustment in natural or human systems to changes in the environment. Adaptation to climate change refers to actions taken by individuals or organizations to adjust to the effects of climate change, including actions that reduce the negative impact of climate change or take advantage of potential new opportunities in response to actual or expected impacts. Examples include building river or coastal dikes, cultivating temperature-shock resistant plants, or changing development pathways.[1,2]

Adaptive capacity (in relation to climate change impacts). "The ability of a system to adjust to climate change (including climate variability and extremes) to moderate potential damages, to take advantage of opportunities, or to cope with the consequences."[1]

Aerosols. A suspension of solid or liquid particles in a gas, with a typical size between 0.01 and 10 micrometers. Aerosols typically reside in the atmosphere for at least several hours and can scatter and absorb radiation. Aerosols have come to be erroneously associated with the propellant used in "aerosol sprays."[1,2]

Afforestation. Planting new forests in areas that historically have not contained forests (in contrast to reforestation).[1,2]

Albedo. The ability of a surface to reflect radiation, often expressed as a percentage. The higher the albedo is, the more radiation reflected. "Snow-covered surfaces have a high albedo, the surface albedo of soils ranges from high to low, and vegetation-covered surfaces and oceans have a low albedo."[1]

Anthropogenic. "Resulting from or produced by human beings."[1]

Atmospheric lifetime. See residence time.

Biofuel. Solid, liquid or gas fuel derived from organic material or combustible oils produced by plants. Examples include ethanol from corn and cellulosic materials, alcohol, wood, and fuels derived from algae.[1]

Biosphere. The part of the earth system in which living organisms exist or that is capable of supporting life, including derived dead organic matter.[1] Carbon reservoirs in the biosphere include organic matter, plants, and soils.[3]

Cap and trade. A market-based program used to control pollution by creating an annual limit of allowable emissions ("cap"). Allowances to emit are distributed among emitters for free, on the basis of historical emissions or other criteria, or sold by the government via auction. Once distributed, the allowances can be traded among emitters, enabling those for which the costs of pollution reduction are high to purchase allowances from those for which such costs are low. The overall cap can be reduced over time, reducing the total amount of allowable emissions. In the United States, cap-and-trade programs are currently in place for SO_2, NO_x, and, locally, for CO_2.

Carbon budget. An accounting of the fluxes (exchanges) among reservoirs, including carbon stored, removed, and added. If fluxes to a reservoir exceed fluxes from the reservoir, the reservoir is a carbon sink.[4]

Carbon capture and storage. An approach to mitigating global warming based on capturing CO_2 from industrial and power-related (large point) sources, moving it to a storage location, and isolating it from the atmosphere for the long term in either biological or geological storage. Biological storage areas include algae lagoons; geological storage areas are underground geologic formations, such as saline aquifers, active or abandoned oil or gas reservoirs, or unminable coal seams.[5,6]

Carbon cycle. The "set of processes such as photosynthesis, respiration, decomposition, and air-sea exchange, by which carbon continuously cycles through various reservoirs, such as the atmosphere, living organisms, soils, and oceans."[7] See also Figure 2-1, Carbon Fluxes and Reservoirs in the Global Carbon Cycle.

Carbon neutral. "Calculating your total climate-damaging carbon emissions, reducing them where possible, and then balancing your remaining emissions, often by purchasing a carbon offset: paying to plant new trees or investing in 'green' technologies such as solar and wind power."[8]

Carbon offset. A measureable reduction, avoidance, or sequestration of GHG emissions relative to business-as-usual activity that can be used to compensate for emissions elsewhere. A key element of offsets is that they must be additional to what otherwise would have been done. In the United States, verified emission reductions are currently bought and sold in a voluntary market. Offsets can be derived from energy efficiency projects, biological sequestration, use of renewable energy in lieu of fossil, and CH_4 projects.[6,9]

Carbon reservoir. "Any natural region or zone, or any artificial holding area, containing an accumulation of carbon or carbon-bearing compounds or having the potential to accumulate such substances."[4]

Carbon sequestration. "The process of increasing the carbon content of a carbon reservoir other than the atmosphere; often used narrowly to refer to increasing the carbon content of carbon [reservoirs] in the biosphere," rather than carbon capture and storage. Biological approaches include increasing carbon storage through changing land use, increasing forest cover, and changing agricultural practices.[4] See also carbon capture and storage.

Carbon sink. A reservoir that accumulates more carbon than it releases. Alternatively, "any process, activity, or mechanism which removes a greenhouse gas or a precursor of a greenhouse gas or aerosol from the atmosphere."[4] The main natural sinks are absorption of CO_2 by the oceans and photosynthesis by plants and algae.[3]

Carbon source. A reservoir that releases more carbon than it stores. Alternatively, "any process, activity, or mechanism which releases a greenhouse gas or a precursor of a greenhouse gas or aerosol into the atmosphere."[4] The main anthropogenic carbon source is the release of CO_2 from fossil fuel use and cement production.[3]

Carbon tax. A levy on the carbon content of fossil fuels that is equivalent to a tax on the CO_2 emitted from their combustion.[10]

Cellulosic biofuels. Renewable fuel "produced from cellulose derived from renewable biomass feedstocks such as corn stover (plant matter left in the field after harvest), switchgrass, wood chips, and other plant or waste matter."[11]

CO_2 equivalent. A quantity that describes the amount of CO_2 that would have the same effect on the radiative balance of the earth (global warming potential), as a "well mixed" GHG, such as CH_4.[4,7]

Cryosphere. The part of the earth system comprising "all snow, ice and frozen ground (including permafrost) on and beneath the surface of the earth and ocean."[1]

Economic potential. "The mitigation potential that takes into account social costs and benefits and social discount rates, assuming that market efficiency is improved by policies and measures, and barriers are removed."[1]

Emission allowance. See emission permit.

Emission permit. Permit (allowance) that allows an entity to emit a specified amount of GHGs, usually within a given period.

Energy balance. The balance between "the amount of solar radiation absorbed, and the amount of longwave radiation emitted by the earth." This balance has been relatively stable over time, meaning the temperature of the earth has been relatively constant.[12]

Fossil fuel. "Carbon-based fuels from fossil hydrocarbon deposits, including coal, peat, oil, and natural gas."[1] See hydrocarbon.

General circulation models (GCMs). Global three-dimensional computer models of the climate system. The atmosphere simulation includes the effects of "reflective and absorptive properties of atmospheric water vapor, [GHG] concentrations, clouds, ...solar heating, ocean temperatures and ice boundaries."[2] The models reproduce clouds, wind, snow, rain, and most weather elements, allowing them to illustrate climate variability. They are sometimes referred to as global climate models. See Figure 2-11, Evolution of Climate Models.

Geothermal power. Energy generated from heat stored in the earth. Applications include direct use of geothermal heat, generating electricity from geothermal heat, and using the shallow ground to heat and cool buildings (heat pumps).[13]

Global warming potential. An index value representing the contribution of one unit mass of a gas to the warming of the earth (through radiative forcing) over a specified period (often 100 years), with CO_2 having a value of 1, and

N_2O, for example, having a value of 298, because it is 298 times more effective at warming the earth than CO_2.[1,4]

Greenhouse effect. The trapping and buildup of heat in the troposphere.[2] GHGs in the atmosphere absorb long wave (infrared) radiation emitted by the earth's surface, atmosphere (due to the same gases), and clouds, and reemit the radiation, maintaining the temperature of the earth. A greater concentration of GHGs in the atmosphere absorbs and reemits more heat to the earth, allowing global temperature to increase (radiative forcing causing an enhanced greenhouse effect).[1]

Greenhouse gas (GHG). Gas in the atmosphere that absorbs and emits radiation within the thermal infrared range; GHGs are essential to maintaining the current temperature of the earth. The primary GHGs include water vapor, carbon dioxide (CO_2), nitrous oxide (N_2O), methane (CH_4) and ozone (O_3). In addition to naturally occurring GHGs, some GHGs are created by human activity.[1] See Table 1-1 in Chapter 1.

Hydrocarbon. An organic compound consisting entirely of hydrogen and carbon.[2] In discussions of climate change, the term is often used to refer to fossil fuels. See fossil fuel.

Hydrologic cycle. "The cycle in which water evaporates from the oceans and the land surface, is carried over the earth in atmospheric circulation as water vapour, condensates to form clouds, precipitates again as rain or snow, is intercepted by trees and vegetation, provides runoff on the land surface, infiltrates into soils, recharges groundwater, discharges into streams, and ultimately, flows out into the oceans, from which it will eventually evaporate again. The various systems involved in the hydrological cycle are usually referred to as hydrological systems."[1]

Hydrosphere. The part of the earth system comprising liquid surface and subterranean water. Carbon reservoirs in the hydrosphere include streams, rivers, lakes, and oceans.[1,3]

(Climate change) Impact assessment. "The practice of identifying and evaluating, in monetary and/or non-monetary terms, the effects of climate change on natural and human systems."[1]

Infrared radiation. Invisible radiation emitted by the earth's surface, atmosphere, and clouds. Infrared radiation is in the part of the electromagnetic spectrum characterized by wavelengths just longer than those of ordinary visible light and shorter than those of microwaves or radio waves. An example is the heat we feel from a glowing coal or light bulb. Infrared radiation is also known as terrestrial or long wave radiation.[2]

Lithosphere. The part of the earth system comprising the "upper layer of the solid earth, both continental and oceanic, which comprises all crustal rocks and the cold, mainly elastic part of the uppermost mantle."[14] Carbon reservoirs in the lithosphere include fossil fuels, sediments, and rocks.[3]

Meridional Overturning Circulation (MOC). An average, large-scale north-south overturning circulation in the oceans. In the Atlantic, this circulation transports relatively warm upper-ocean waters northward and relatively cold waters southward (the Gulf Stream is a part of this). In contrast to the concept of thermohaline circulation, MOC can also include shallow, wind-driven overturning in the tropics and subtropics.[1]

Metric ton. A measurement of mass equal to 1,000 kilograms, or 2,204.62 pounds.

Mitigation. Actions taken by individuals or organizations to reduce GHG emissions or enhance carbon sinks to minimize the effects of GHGs on global climate change.[4,7]

Monitoring. A process of tracking progress toward a goal or objective, specifically the implementation of an adaptation strategy and its components.[15]

No-till farming. A way of growing crops that "does away with the mechanical churning of the soil between crops. Rather than being plowed under, agricultural wastes are left on top of the soil, where they decompose more slowly."[16]

Post-Special Report on Emission Scenarios scenarios. Emission scenarios published after the IPCC SRES.[1] See Special Report on Emission Scenarios. See also Figure 2-9, Global GHG Emissions for 2000 and Projected Emissions for 2030 using SRES and Post-SRES Scenarios.

Proactive adaptation. Adaptation action "that takes place before impacts of climate change are observed."[1] Also known as anticipatory adaptation.

Radiative forcing. "Radiative forcing is the change in the net vertical irradiance (expressed in Watts per square meter) at the tropopause [the boundary of the troposphere and stratosphere] due to an internal change or a change in the external forcing of the climate system."[2]

Reactive adaptation. Adaptation action that takes place in response to impacts of climate change.

Residence time. The average time an individual atom or molecule remains in a reservoir. This term is used to refer to the time a GHG molecule remains in the atmosphere.[2]

Second generation biofuels. Biofuels produced from cellulose, available from nonfood crops and waste biomass (wood byproducts, straw) in contrast to first-generation biofuels, which are produced from sugar and starch crops (sugarcane, corn, soybean).[17]

Sensitivity analysis. "Simulation analysis in which key quantitative assumptions and computations (underlying a decision, estimate, or project) are changed systematically to assess their effect on the final outcome. Employed commonly in evaluation of the overall risk or in identification of critical factors, it attempts to predict alternative outcomes of the same course of action. Also called what-if analysis."[18]

Sensitivity. "The degree to which a system is affected, either adversely or beneficially, by climate variability or change. The effect may be direct (e.g., a change in crop yield in response to a change in the mean, range or variability of temperature) or indirect (e.g., damages caused by an increase in the frequency of coastal flooding due to sea-level rise)."[1] This concept of sensitivity is not to be confused with climate sensitivity, which refers to "the equilibrium change in the annual mean global surface temperature following a doubling of the atmospheric equivalent carbon dioxide concentration."[1]

Social Cost of Carbon (SCC). An estimate of the cost of climate change damages to economies and societies, measured as the discounted value (in $/ton CO_2) of "annual net losses from impacts triggered by an additional ton of carbon emitted today."[7]

Special Report on Emissions Scenarios. Published by the IPCC in 2000, outlines scenarios that identify demographic change, social and economic development, and the rate and direction of technological change as the main driving forces of future GHG trajectories. SRES scenarios that have a "similar demographic, societal, economic, and technical-change storyline" are grouped into families (A1, A2, B1, and B2).[19] See: Figure 2-8, SRES Qualitative Storylines for Modeling Scenarios; Figure 2-9, Global GHG Emissions for 2000 and Projected Emissions for 2030 using SRES and Post-SRES Scenarios; and Figure 2-10, Global Averages of Surface Warming for SRES Scenarios.

Stratosphere. The second layer of the atmosphere between the troposphere and mesosphere between about 10 km to 50 km in altitude. The temperature of the upper stratosphere increases with height due to absorption of solar radiation by O_3. The temperature of the lower stratosphere can increase, be isothermal, or decrease with altitude depending on latitude and season.[2]

Terrestrial radiation. See infrared radiation.

Thermohaline circulation. "Large-scale density-driven circulation in the ocean, caused by differences in temperature and salinity."[2] See Figure 2-6, Thermohaline Circulation of Global Ocean System.

Transpiration. "The evaporation of water vapour from the surfaces of leaves through stomata."[20]

Troposphere. The lowest region of the atmosphere, from the surface to about 10 km in altitude, characterized by decreasing temperature with increasing altitude. This is where clouds and "weather" occur.[1]

Vulnerability. The degree to which systems are susceptible to, and unable to cope with, adverse impacts of climate change.[1]

Warming potential. See global warming potential.

---

[1] IPCC, "Glossary" in *Climate Change 2007: Synthesis Report. Contribution of Working Groups I, II, and III to the Fourth Assessment Report of the Intergovernmental Panel on Climate Change*, Cambridge University Press, Cambridge, United Kingdom, New York, NY, USA, www.ipcc.ch/ipccreports/ar4-syr.htm.

[2] EPA, *Glossary of Climate Change Terms*, www.epa.gov/climatechange/glossary.html.

[3] National Aeronautics and Space Administration Earth Observatory, *The Carbon Cycle: The Human Role*, 2001, earthobservatory.nasa.gov/Library/CarbonCycle/carbon_cycle4.html.

[4] CCSP and the Subcommittee on Global Change Research, *The First State of the Carbon Cycle Report (SOCCR): The North American Carbon Budget and Implications for the Global Carbon Cycle*, Synthesis and Assessment Product 2.2, [King, A.W. et al. (eds.)], Glossary, NOAA, National Climatic Data Center, Asheville, NC., www.climatescience.gov/Library/sap/sap2-2/final-report/sap2-2-final-glossary.pdf.

[5] IPCC, *IPCC Special Report on Carbon Dioxide Capture and Storage*. Prepared by Working Group III of the Intergovernmental Panel on Climate Change [Metz, B., O. Davidson, H. C. de Coninck, M. Loos, and L. A. Meyer (eds.)], Cambridge University Press, Cambridge, United Kingdom and New York, NY, USA, 442 pp., 2005, www.ipcc.ch/pdf/special-reports/srccs/srccs_wholereport.pdf.

[6] GAO, *Carbon Offsets: The U.S. Voluntary Market Is Growing, but Quality Assurance Poses Challenges for Market Participants*, GAO-08-1048, September 26, 2008, www.gao.gov/htext/d081048.html.

[7] IPCC, *Climate Change 2007: Mitigation of Climate Change. Working Group III Contribution to the Fourth Assessment Report of the Intergovernmental Panel on Climate Change*, Cambridge University Press, Cambridge, United Kingdom, and New York, NY, USA, www.ipcc.ch/pdf/glossary/ar4-wg3.pdf

[8] Oxford University Press, *The New Oxford American Dictionary*, Second Edition, Erin McKean (Ed.), 2051 pp., May 2005, ISBN 0-19-517077-6.

[9] Jonathan L. Ramseur, *The Role of Offsets in a Greenhouse Gas Emissions Cap-and-Trade Program: Potential Benefits and Concerts*, Congressional Research Service, CRS Report for Congress, RL34436, April 4, 2008, fpc.state.gov/documents/organization/103675.pdf.

[10] IPCC, "Glossary" in *Climate Change 2001: Synthesis Report. A Contribution of Working Groups I, II, and III to the Third Assessment Report of the Intergovernmental Panel on Climate Change*, Cambridge University Press, Cambridge, United Kingdom, and New York, NY, USA, www.ipcc.ch/pdf/glossary/tar-ipcc-terms-en.pdf.

[11] Tom Capehart, *Cellulosic Biofuels: Analysis of Policy Issues for Congress*, Congressional Research Service Report for Congress RL34738, November 7, 2008, fpc.state.gov/documents/organization/112413.pdf.

[12] NASA Earth Observatory, *The Earth's Energy Balance*, earthobservatory.nasa.gov/Experiments/PlanetEarthScience/GlobalWarming/GW_Movie2.php.

[13] NREL, *Geothermal Energy Basics, Learning about Renewable Energy*, www.nrel.gov/learning/re_geothermal.html.

[14] IPCC, "Glossary" in *Climate Change 2007: The Physical Science Basis. Working Group I Contribution to the Fourth Assessment Report of the Intergovernmental Panel on Climate Change*, Cambridge University Press, Cambridge, United Kingdom, and New York, NY, USA, www.ipcc.ch/ipccreports/ar4-wg1.htm.

[15] Rosa T. Perez, et al., "Continuing the Adaptation Process," in *Policy Frameworks for Climate Change: Developing Strategies, Policies, and Measures*, United Nations Development Program, and Global Environmental Facility, www.undp.org/gef/undp-gef_publications/publications/apf%20technical%20paper09.pdf.

[16] Lisa Raffensperger, *A Fresh Green for No-Till Farming*, World Resources Institute, 2008, earthtrends.wri.org/updates/node/286.

[17] EPA, *Economics of Biofuels*, yosemite.epa.gov/EE/epa/eed.nsf/pages/Biofuels.html.

[18] Business Dictionary, *Sensitivity analysis,* WebFinance Inc., www.businessdictionary.com/definition/sensitivity-analysis.html.

[19] CCSP, 2008, *Analyses of the effects of global change on human health and welfare and human systems*, A Report by the U.S. Climate Change Science Program and the Subcommittee on Global Change Research, Glossary and Acronyms, [Gamble, J.L. (ed.), K.L. Ebi, F.G. Sussman, T.J. Wilbanks, (Authors)], U.S. EPA, Washington, DC, USA, www.climatescience.gov/Library/sap/sap4-6/final-report/.

[20] IPCC, "Glossary" in *Climate Change 2007: Impacts, Adaptation, and Vulnerability. Working Group II Contribution to the Fourth Assessment Report of the Intergovernmental Panel on Climate Change,* Cambridge University Press, Cambridge, United Kingdom, and New York, NY, USA, www.ipcc.ch/ipccreports/ar4-wg2.htm.

Appendix F
Abbreviations

AR4	fourth assessment report
ASHRAE	American Society of Heating, Refrigerating and Air Conditioning Engineers
BLM	Bureau of Land Management
CAA	Clean Air Act
CAFE	Corporate Average Fuel Economy
CCAR	California Climate Action Registry
CCCSTI	Cabinet-level Committee on Climate Change Science and Technology Integration
CCSP	Climate Change Science Program
CCTP	Climate Change Technology Program
CCX	Chicago Climate Exchange
CDC	Centers for Disease Control and Prevention
CEA	Council of Economic Advisors
CEQ	Council on Environmental Quality
CFL	compact fluorescent light
CIA	Central Intelligence Agency
CNG	compressed natural gas
CO_2-eq	carbon dioxide equivalent
COP	Conference of the Parties
CPG	Comprehensive Procurement Guidelines
CRS	Congressional Research Service
DC	District of Columbia
DHS	Department of Homeland Security
DOC	Department of Commerce
DoD	Department of Defense
DOE	Department of Energy
DOI	Department of the Interior
DOS	Department of State
DOT	Department of Transportation
DPC	Domestic Policy Council
DS	Deputy Secretary
E85	ethanol 85 percent

EIA	Energy Information Administration
EISA	Energy Independence and Security Act of 2007
EO	executive order
EPA	Environmental Protection Agency
EPAct	Energy Policy Act
EU	European Union
FBI	Federal Bureau of Investigation
FEMP	Federal Energy Management Program
FTC	Federal Trade Commission
FY	fiscal year
G8	the Group of Eight (U.S., Germany, France, Canada, Italy, Japan, Russia, and the United Kingdom)
GAO	Government Accountability Office
GCM	general circulation model
GDP	gross domestic product
GHG	greenhouse gas
GOOS	good on one side
GSA	General Services Administration
HHS	Department of Health and Human Services
HUD	Housing and Urban Development
ICE	internal combustion engine
ICLEI	International Council for Local Environmental Initiatives
IPCC	Intergovernmental Panel on Climate Change
IRI	International Research Institute for Climate and Society
LEED	Leadership in Energy and Environmental Design
MOC	meridional overturing circulation
NAS	National Academy of Sciences
NASA	National Aeronautics and Space Administration
NEC	National Economic Council
NIH	National Institutes of Health
NIST	National Institute of Standards and Technology
NOAA	National Oceanic and Atmospheric Administration
NPS	National Park Service
NRC	National Research Council
NREL	National Renewable Energy Lab
NSC	National Security Council
NSF	National Science Foundation

OECD	Organization for Economic Cooperation and Development
OFDA	Office of Foreign Disaster Assistance
OMB	Office of Management and Budget
OSTP	Office of Science and Technology Policy
PDI	power dissipation index
PRM	Bureau of Population, Refugees and Migration
R&D	research and development
RECLAIM	Regional Clean Air Incentives Market
RGGI	Regional Greenhouse Gas Initiative
SAP	synthesis and assessment products
SCC	social cost of carbon
SPM	Summary for Policy Makers
SRES	Special Report on Emissions Scenarios
UK	United Kingdom
UN	United Nations
UNFCCC	United Nations Framework Convention on Climate Change
U.S.	United States
US	Under Secretary
USAID	United States Agency for International Development
USDA	United States Department of Agriculture
USFS	United States Forest Service
USFWS	United States Fish and Wildlife Service
USGCRP	U.S. Global Change Research Program
USGS	United States Geological Survey
USPS	United States Postal Services
UV	ultraviolet
WCI	Western Climate Initiative

Units

°C	degrees Celsius
°F	degrees Fahrenheit
cm	centimeter
gge	gasoline gallon equivalent MPG per mile
Gt	gigaton (billion tons)
kg	kilogram
km	kilometer
kWh	kilowatt-hour

lb	pound
mm	millimeter
mpg	miles per gallon
ppb	parts per billion
ppm	parts per million
ppt	parts per trillion
Mt	megaton (million tons)
W/m^2	^{w}atts per meter square

CHEMICAL FORMULAS

Ar	argon
CFC	chlorofluorocarbon
CH_4	methane
CO_2	carbon dioxide
HFC	hydrofluorocarbon
N_2	nitrogen
N_2O	nitrous oxide
NO_x	nitrogen oxide (generic)
O_2	oxygen
O_3	ozone
PFC	perfluorocarbon
SF_6	sulfur hexafluoride
SO_2	sulfur dioxide
SO_x	sulfur oxide

Index